MW00814764

Modern Magnetic and Spintronic Materials

NATO Science for Peace and Security Series

This Series presents the results of scientific activities supported through the framework of the NATO Science for Peace and Security (SPS) Programme.

The NATO SPS Programme enhances security-related civil science and technology to address emerging security challenges and their impacts on international security. It connects scientists, experts and officials from Alliance and Partner nations to work together to address common challenges. The SPS Programme provides funding and expert advice for security-relevant activities in the form of Multi-Year Projects (MYP), Advanced Research Workshops (ARW), Advanced Training Courses (ATC), and Advanced Study Institutes (ASI). The NATO SPS Series collects the results of practical activities and meetings, including:

Multi-Year Projects (MYP): Grants to collaborate on multi-year R&D and capacity building projects that result in new civil science advancements with practical application in the security and defence fields.

Advanced Research Workshops: Advanced-level workshops that provide a platform for experts and scientists to share their experience and knowledge of security-related topics in order to promote follow-on activities like Multi-Year Projects.

Advanced Training Courses: Designed to enable specialists in NATO countries to share their security-related expertise in one of the SPS Key Priority areas. An ATC is not intended to be lecture-driven, but to be intensive and interactive in nature.

Advanced Study Institutes: High-level tutorial courses that communicate the latest developments in subjects relevant to NATO to an advanced-level audience.

The observations and recommendations made at the meetings, as well as the contents of the volumes in the Series reflect the views of participants and contributors only, and do not necessarily reflect NATO views or policy.

The series is published by IOS Press, Amsterdam, and Springer, Dordrecht, in partnership with the NATO SPS Programme.

Sub-Series

A.	Chemistry and Biology	Springer
B.	Physics and Biophysics	Springer
C.	Environmental Security	Springer
D.	Information and Communication Security	IOS Press
E.	Human and Societal Dynamics	IOS Press

- http://www.nato.int/science
- http://www.springer.com
- http://www.iospress.nl

Series B: Physics and Biophysics

Modern Magnetic and Spintronic Materials

Properties and Applications

edited by

Andreas Kaidatzis
National Centre for Scientific Research "Demokritos"
Institute of Nanoscience and Nanotechnology
Athens, Greece

Serhii Sidorenko
Metals Physics Department
National Technical University of Ukraine
Kiev, Ukraine

Igor Vladymyrskyi
Metals Physics Department
National Technical University of Ukraine
Kiev, Ukraine

and

Dimitrios Niarchos
National Centre for Scientific Research "Demokritos"
Institute of Nanoscience and Nanotechnology
Athens, Greece

 Springer

Published in Cooperation with NATO Emerging Security Challenges Division

Results from the NATO Advanced Training Course on Spintronics Radar Detectors
Athens, Greece
14–18 October 2019

ISBN 978-94-024-2036-4 (PB)
ISBN 978-94-024-2033-3 (HB)
ISBN 978-94-024-2034-0 (eBook)
https://doi.org/10.1007/978-94-024-2034-0

Published by Springer,
P.O. Box 17, 3300 AA Dordrecht, The Netherlands.

www.springer.com

Printed on acid-free paper

Preface

Materials have been technology enablers throughout human prehistory and history, as manifested by the designation of various time periods according to the materials humans used, e.g. Bronze or Iron Age.

Magnetic materials are ubiquitous in modern technology, as permanent magnets in electric motors, power generators, sensors and actuators; as soft magnets in alternating current or high frequency applications; and as hard magnetic films in data storage applications. The past three decades after the advent of spin- and magnon-electronics – dubbed "spintronics" and "magnonics" – magnetic materials are also employed in non-volatile memories and information processing applications. Medical technology has also been benefited by magnetic materials – especially magnetic nanoparticles – for therapy and diagnostics methods, e.g. in cancer treatment using magnetic hyperthermia or in biomedical imaging using contrast enhancing agents.

All of the above-mentioned applications rely on the magnetic properties of the materials used. These properties depend on intrinsic and extrinsic material parameters. The former are related to the actual elements used and their properties, e.g. atomic magnetic moment and exchange interaction between atoms. The latter are related to the structural and microstructural properties of the materials, e.g. crystal structure, grain size and grain boundary phases.

This volume is dealing with state-of-the-art applications of magnetic and spintronic materials and the reader will be exposed to various approaches of the related Materials Science and Engineering. Phenomena and processes appearing at the nanoscale are of particular importance for modern magnetic and spintronic materials and a great part of the content is dealing with such topics. The chapters are "stand-alone", allowing a non-serial study of the volume.

Some of the contributions to this volume were presented as lectures at the Advanced Training Course "Spintronics Radar Detectors" (14–18 October 2019,

Athens, Greece), an activity supported by the NATO Science for Peace and Security Programme. The editors would like to acknowledge the valuable assistance of Mr. Arsen Hafarov from National Technical University of Ukraine "Igor Sikorsky Kyiv Polytechnic Institute" during volume preparation.

Athens, Greece   Andreas Kaidatzis
Kiev, Ukraine   Serhii Sidorenko
Kiev, Ukraine   Igor Vladymyrskyi
Athens, Greece   Dimitrios Niarchos

Contents

Contributors

Antonis Alexandridis Institute of Informatics and Telecommunications, NCSR "Demokritos", Athens, Greece

Petro Yu. Artemchuk Faculty of Radio Physics, Electronics and Computer Systems, Taras Shevchenko National University of Kyiv, Kyiv, Ukraine

Igor O. Fritsky Faculty of Chemistry, Taras Shevchenko National University of Kyiv, Kyiv, Ukraine

Szilvia Gulyas Faculty of Science and Technology, Department of Solid State Physics, University of Debrecen, Debrecen, Hungary
Doctoral School of Physics, University of Debrecen, Debrecen, Hungary

Il'ya A. Gural'skiy Faculty of Chemistry, Taras Shevchenko National University of Kyiv, Kyiv, Ukraine

Arsen Hafarov Metal Physics Department, National Technical University of Ukraine "Igor Sikorsky Kyiv Polytechnic Institute", Kyiv, Ukraine

Gábor L. Katona Faculty of Science and Technology, Department of Solid State Physics, University of Debrecen, Debrecen, Hungary

Michał Krupiński Institute of Nuclear Physics Polish Academy of Sciences, Krakow, Poland

Olesia I. Kucheriv Faculty of Chemistry, Taras Shevchenko National University of Kyiv, Kyiv, Ukraine

Vilen L. Launets Faculty of Radio Physics, Electronics and Computer Systems, Taras Shevchenko National University of Kyiv, Kyiv, Ukraine

Denys Makarov Helmholtz-Zentrum Dresden-Rossendorf e.V., Institute of Ion Beam Physics and Materials Research, Dresden, Germany

Iurii Makogon Metals Physics Department, National Technical University of Ukraine "Igor Sikorsky Kyiv Polytechnic Institute", Kyiv, Ukraine

Pavlo Makushko Metals Physics Department, National Technical University of Ukraine "Igor Sikorsky Kyiv Polytechnic Institute", Kyiv, Ukraine

Euthimios Manios Institute of Nanoscience and Nanotechnology, NCSR "Demokritos", Athens, Greece

Marta Marszalek Institute of Nuclear Physics Polish Academy of Sciences, Krakow, Poland

Viktor V. Oliynyk Faculty of Radio Physics, Electronics and Computer Systems, Taras Shevchenko National University of Kyiv, Kyiv, Ukraine

Michael Pissas Institute of Nanoscience and Nanotechnology, NCSR "Demokritos", Athens, Greece

Oleksandr V. Prokopenko Faculty of Radio Physics, Electronics and Computer Systems, Taras Shevchenko National University of Kyiv, Kyiv, Ukraine

Mark Shamis Metals Physics Department, National Technical University of Ukraine "Igor Sikorsky Kyiv Polytechnic Institute", Kyiv, Ukraine

Serhii Sidorenko Metal Physics Department, National Technical University of Ukraine "Igor Sikorsky Kyiv Polytechnic Institute", Kyiv, Ukraine

Ilias Tsiachristos Institute of Nanoscience and Nanotechnology, NCSR "Demokritos", Athens, Greece

Erene Varouti Institute of Nanoscience and Nanotechnology, NCSR "Demokritos", Athens, Greece

Tetiana Verbytska Metals Physics Department, National Technical University of Ukraine "Igor Sikorsky Kyiv Polytechnic Institute", Kyiv, Ukraine

Igor Vladymyrskyi Metal Physics Department, National Technical University of Ukraine "Igor Sikorsky Kyiv Polytechnic Institute", Kyiv, Ukraine

Yevhen Zabila Institute of Nuclear Physics Polish Academy of Sciences, Krakow, Poland

Volodymyr V. Zagorodnii Faculty of Radio Physics, Electronics and Computer Systems, Taras Shevchenko National University of Kyiv, Kyiv, Ukraine

About the Editors

Andreas Kaidatzis Institute of Nanoscience and Nanotechnology, National Centre for Scientific Research "Demokritos", Athens, Greece
e-mail: a.kaidatzis@inn.demokritos.gr
Dr. Kaidatzis holds a first degree in Physics (2003) and a postgraduate degree in Materials Physics (2005), obtained from the Aristotle University of Thessaloniki, Greece, and a doctoral degree in Solid State Physics obtained from the University of Paris 11 (2008) as a Marie-Curie Fellow. He has worked as Postdoctoral Researcher at the Institute of Microelectronics of NCSR "Demokritos" (2010–2011) and at the Institute of Microelectronics of Madrid (2011–2013) in the framework of an Individual Marie-Curie Fellowship. His work was focused on the study of magnetic thin films and nanostructures using magnetic force microscopy. Since May 2013 he is working as a Research Associate at the Institute of Nanoscience and Nanotechnology of NCSR "Demokritos", studying magnetic and spintronic materials for memories, data recording, sensors, and permanent magnets applications. Dr. Kaidatzis has participated in several international research projects, granted from the European Commission, the Spanish research council (CSIC) and the Hellenic General Secretariat for Research and Technology. He has published his work in 18 peer-reviewed journals and presented more than 50 abstracts in national and international conferences.

Serhii Sidorenko Metals Physics Department, National Technical University of Ukraine "Igor Sikorsky Kyiv Polytechnic Institute", Kyiv, Ukraine
e-mail: sidorenko@kpi.ua
Prof. Dr. Sidorenko graduated from the Moscow Institute of Steel and Alloys (1971). In 1987 he defended his doctoral thesis. Research activities of Prof. Sidorenko are linked with advanced technologies on the basis of theoretical and

---

The editors were co-organizers and lecturers at the Advanced Training Course "Spintronics Radar Detectors", held from 14th to 18th October 2019 in Athens (Greece). This activity was supported by the NATO "Science for Peace and Security" Programme.

experimental investigations of diffusion processes, phase formation and other phenomena in different metallic materials. He has authored over 300 papers in area of physical materials science. Under his leadership, 24 PhD theses were defended.

Igor Vladymyrskyi Metals Physics Department, National Technical University of Ukraine "Igor Sikorsky Kyiv Polytechnic Institute", Kyiv, Ukraine

e-mail: vladymyrskyi@kpm.kpi.ua

Dr. Vladymyrskyi is an Associate Professor of Metals Physics Department of National Technical University of Ukraine "Igor Sikorsky Kyiv Polytechnic Institute". He received his PhD degree from the same University in 2014. Research interests of Dr. Vladymyrskyi focus on structural phase transitions in nanoscale metallic layered thin films; diffusion and diffusion-controlled phenomena in these materials; and their structural, magnetic and electrical properties. Recent works are devoted to thermally induced diffusion formation of ferromagnetic $L1_0$ ordered thin films – e.g. $L1_0$-FePt, $L1_0$-FePd, $L1_0$-MnAl – promising materials for applications as magnetic recording medium with ultrahigh recording density or ferromagnetic electrodes of spintronic devices with an excellent functionality. Dr. Vladymyrskyi has coauthored about 30 papers in international peer-reviewed journals. He is a co-leader of a joint project with Augsburg University (Germany) called "Cold homogenization of Fe/Pt based layered thin films induced by diffusion processes" funded by German Research Foundation (DFG).

Dimitrios Niarchos National Centre for Scientific Research "Demokritos", Institute of Nanoscience and Nanotechnology, Athens, Greece

e-mail: d.niarchos@inn.demokritos.gr

Prof. Dr. Niarchos obtained his BSc and PhD in Physics-Materials Science from National Kapodestrian University of Athens, Greece, in 1972 and 1980, respectively. He holds also an MBA of R&D from Loyola University in 1985. From 1978 to 1981 he was IAEC Distinguished Postdoctoral Researcher at Argonne National Lab, and from 1981 to 1985 he was Assistant Professor at the Illinois Institute of Technology, Chicago, USA. He then moved to the Institute of Materials Science of the NCSR "Demokritos" in 1985 as Senior Researcher. He was promoted to Director of Research (Grade A) in 1992. From 1994 to 1999 he served as the Director of Institute of Materials Science of the NCSR "Demokritos", and from 1996 to 1999 he was Vice President of the NCSR "Demokritos". From 2005 to 2010 he was Director and President of the Board of the NCSR "Demokritos". He is studying magnetic materials for novel permanent magnets, ultrahigh recording media, magnetic memories and magnetic micro-nanolaminated structures for UHF DC-DC converters. He is the author and co-author of more than 490 publications with more than 8000 citations and has managed more than 35 National and EU projects with a budget of approximately 12 MEURO. He had served as advisor for the Greek Government, NATO and the EC. He holds six patents related to nanomaterials, thermoelectrics and high entropy alloys for permanent magnets.

Chapter 1
Detection of Microwave and Terahertz-Frequency Signals in Spintronic Nanostructures

Petro Yu. Artemchuk and Oleksandr V. Prokopenko

Abstract We consider recent theoretical and experimental results related to the properties of microwave and terahertz-frequency (TF) detectors based on multilayered spintronic nanostructures, so-called spintronic diodes (SDs). In such a diode, input ac current excites magnetization oscillations in a free magnetic layer of spintronic nano-structure that gives rise to the ac variations of nanostructure's magnetoresistance with frequency that is close to the input signal frequency. The mixing of ac current and ac magnetoresistance oscillations produces an output low-frequency/dc voltage across the SD. In the other type of detector based on a magnetic/heavy metal bilayer structure, the detector's output voltage is generated in the metal layer due to the conversion of excited ac magnetization oscillations to the low-frequency/dc voltage via the inverse spin Hall effect and the spin pumping effect. It was shown theoretically and experimentally that such spintronic detectors can be ultra-sensitive to microwave and TF radiation that makes them promising for a wide range of applications. In the paper we consider typical properties of such detectors based on different types of spintronic nanostructures working in distinct operation regimes and analyze their benefits and drawbacks, and also review their prospective applications.

Keywords Microwave detector · Terahertz-frequency detector · Spintronic diode · Spin-torque diode effect · Inverse spin Hall effect · Spin pumping effect

1.1 Introduction

Detection of ac signals and measurement of their characteristics (power, spectrum, etc.) is one of the main tasks required for proper subsequent signal processing and storage the signal-coded information in modern electronics [35, 45]. Many

P. Y. Artemchuk · O. V. Prokopenko (✉)
Faculty of Radio Physics, Electronics and Computer Systems, Taras Shevchenko National University of Kyiv, Kyiv, Ukraine
e-mail: ovp@univ.kiev.ua

© Springer Nature B.V. 2020
A. Kaidatzis et al. (eds.), *Modern Magnetic and Spintronic Materials*, NATO Science for Peace and Security Series B: Physics and Biophysics,
https://doi.org/10.1007/978-94-024-2034-0_1

1

promising applications of microwave and terahertz-frequency (TF) electromagnetic signals in medicine, general and military security, communications, material science, space technologies etc. require compact, easy tuned and reliable detectors of coherent and non-monochromatic low-power ac signals [9, 16, 24, 35, 45, 50, 56].

Nowadays commercial signal detectors or their prototypes are mainly fabricated on the basis of semiconductor materials (e.g., see Herotek Inc.) or superconducting Josephson junctions [18, 29, 44, 48]. However, it should be noted that such detectors have several disadvantages. Josephson junctions can work only at cryogenic temperatures, which is unacceptable for large-scale everyday applications. Commercial microwave signal detectors based on semiconductor diodes provide rather low volt-watt sensitivity (responsivity) about 0.5 V/W for unbiased detectors and about 3.8 V/W for dc biased devices (e.g., see products of Herotek Inc.). The performance of semiconductor detectors in the TF band becomes even worse, their volt-watt sensitivity decreases, noise characteristics deteriorate, while the minimum detectable signal power increases [9].

To overcome the limitations of superconducting and semiconductor devices, there is a temptation to use alternative spintronic technologies to detect microwave and TF electromagnetic signals [2, 44]. Note, although such TF detectors may have average or low performance at the moment in comparison to semiconductor devices, their technical parameters could be substantially improved in time, which opens the possible ways for their wide applications in the future. There are two types of such devices, so-called spintronic diodes (SDs). In a SD based on a magnetoresistive junction, input ac current excites the junction magnetoresistance oscillations that being mixed with the current produces an output dc voltage signal [17, 37, 42, 57]. The principles of operation of such detectors, so-called spin-torque detectors, are similar to the principles of operation of spin-torque nano-oscillators [15, 23, 25, 26, 36, 51]; the essential difference between them is that the nano-oscillator generates an output ac signal, when a bias dc signal is subjected to it, while in the detector an input ac signal rectifies and a corresponding output dc signal is produced [17, 37, 42, 57]. In the second type of detector, so-called the spin Hall detector, an output dc signal is generated in a metal layer of magnetic / metal bilayer nanostructure due to the inverse spin Hall and spin pumping effects [14, 33, 49]. The principle of operation of such a device is inversed to the operation principle of a spin Hall nano-oscillator [7, 14, 49].

Below we will briefly consider both these types of detectors, describe their possible operation regimes and explain the main physical phenomena utilized in these devices. Then we will illustrate several typical applications of discussed spintronic detectors in microwave and TF band.

1.2   Basic Phenomena in Spintronic Detectors

The SDs mainly utilize several partially related effects: the spin transfer torque effect, the tunneling magnetoresistance effect, the spin-torque diode effect, the spin Hall effects, and the spin pumping effect.

The Spin-Transfer Torque Effect This effect was theoretically predicted by J. Slonczewski and L. Berger [4, 52] and then experimentally observed and thoroughly studied in many research groups [15, 17, 20, 21, 23, 25, 26, 30, 46, 57, 58]. It allows one to alter the direction of the magnetization vector **M** in nano-scale magnetic systems, if a spin-polarized or pure spin current is subjected to a free magnetic layer (FL) of a system [51]. Hence, in accordance to the spin-transfer torque effect, the applied current can excite magnetization oscillations in the FL [15, 20, 23, 25, 26, 30, 46], or can provide the magnetization switching [21, 58]. The first (generation) regime is suitable for the development of nano-scale signal sources, so-called the spin-torque nano-oscillators [23, 38, 51], and detectors [17, 37, 39, 42, 57]. The other (switching) regime is convenient for the development of fast and reliable memory cells.

The spin-transfer torque effect manifests itself as an additional torque, which do influence on magnetization of the SD's magnetic layer. In the case when the direction of spin polarization of the current $I(t)$ can be considered as completely "fixed", the spin-transfer torque that acts on the magnetization **M** can be written as [51, 59].

$$\frac{\partial \mathbf{M}}{\partial t} = \frac{\sigma(\beta)I(t)}{M_s} [\mathbf{M} \times [\mathbf{M} \times \mathbf{p}]] , \qquad (1.1)$$

where $M_s = |\mathbf{M}|$ is the saturation magnetization of the FL, **p** is the unit vector in the direction of the spin polarization, and $\sigma(\beta)$ is the current-torque proportionality coefficient that is equal to

$$\sigma(\beta) = \frac{\sigma_\perp}{1 + \eta^2 \cos \beta}, \sigma_\perp = \left(\frac{\gamma \hbar}{2e}\right) \frac{\eta}{M_s V} . \qquad (1.2)$$

Here η is the spin-polarization efficiency of the current, $\gamma \approx 2\pi \cdot 28\text{GHz/T}$ is the modulus of the gyromagnetic ratio, $\hbar$ is the reduced Planck constant, e is the modulus of the electron charge, V is the volume of the FL, and $\beta = \arccos(\mathbf{m} \cdot \mathbf{p})$ is the angle between the unit magnetization vector $\mathbf{m} = \mathbf{M}/M_s$ and the unit vector **p**.

The Tunneling Magnetoresistance Effect In magnetic nanostructures with several magnetic layers separated by a tunneling barrier (e.g., in magnetic tunnel junctions), it appears that nanostructure resistance depends on the angle β between the magnetizations of these magnetic layers. In the simplest case of unbiased symmetric magnetoresistive junction this dependence can be written as [42, 60].

$$R(\beta) = \frac{R_\perp}{1 + \eta^2 \cos \beta}, \qquad (1.3)$$

where $R_\perp$ is the junction resistance for the case $\beta = \pi/2$. If the junction area S, the junction material and fabrication technology characterized by the resistance-area product RA are known, characteristic resistance $R_\perp$ can be estimated as $R_\perp = RA/S$. It also follows from Eq. (1.3) that the junction resistance can vary in the range from $R_P = R_\perp/(1 + \eta^2)$ ("parallel" magnetic state) to $R_{AP} = R_\perp/(1 - \eta^2)$ ("antiparallel" magnetic state), while the tunneling magnetoresistance ratio can be defined as $\text{TMR} = 2\eta^2/(1 - \eta^2)$.

In a single-magnetic-layer SD, for instance in an antiferromagnetic tunnel junction Pt/antiferromagnet/MgO/Pt discussed below, the junction magnetoresistance can also depend on the angle between the magnetization direction (or the direction of spin polarization of an applied current) and some direction defined by a crystalline structure of the used magnetic layer. This effect called the tunneling anisotropic magnetoresistance can be utilized for the generation of current-driven resistance oscillations in nanostructures with a single magnetic layer [34].

The Spin-Torque Diode Effect It is a rectification effect of the input ac current $I(t)$ in a magnetoresistive junction, which is commonly observed when the frequency f of the current is close to the ferromagnetic resonance (FMR) frequency f_0 of the FL [17, 37, 42, 57]. In this case the mixing of the junction magnetoresistance oscillations $R(t) \equiv R(\beta(t))$ (at frequency f_0) with the oscillations of input ac current $I(t)$ (at frequency $f \approx f_0$) produces a large enough output dc (or very low frequency) voltage

$$U_{dc} = \langle I(t)R(t) \rangle \tag{1.4}$$

across the junction. In Eq. (1.4) expression $\langle I(t)R(t) \rangle$ denotes averaging $I(t)R(t)$ over the period of current oscillations $1/f$.

One should note that the formulation of the spin-torque diode effect presented above implies that the frequencies f and f_0 are almost equal (otherwise the diode's output mixing signal might be too small to measure). However, the closeness of frequencies f and f_0 is not an absolutely necessary condition to observe the appearance of an output dc voltage in a magnetoresistive junction under the action of ac signal. It will be shown below that there is a SD operation regime, where the diode works as a non-resonance device, but still provides a large enough output rectified voltage generated by an incident ac signal [37, 39, 42].

The Spin Hall Effects In accordance to the direct spin Hall effect, moving electrons with spins "up" and "down" are separating in a thin metal layer with a strong spin-orbit interaction and accumulating on the opposite sides of the layer [49]. This accumulation of spins with different "directions" on the opposite boundaries of the metal layer creates a non-equilibrium spin state that can be treated as an appearance of a transverse spin current (its direction is perpendicular to the direction of an electric current). This spin current then can act on a magnetic state of some magnetic layer attached to the layer of metal. In particular this effect can be used for the excitation of magnetization oscillations in bilayer magnetic/metal systems, so-called the spin Hall nano-oscillators [7, 8].

However, the inverse phenomenon, the so-called inverse spin Hall effect [14, 49] is also possible. This effect manifest itself as a transformation of spin current flowing from a magnetic layer to an electric current in a metal layer at the interface magnetic/metal of bilayer nanostructure.

The Spin Pumping Effect This last effect provides the generation of a spin current in magnetic layer with rotating magnetization vector [14, 33]. In this case the generated spin current can be estimated as

$$\mathbf{j}_{SP} = \frac{\hbar}{4\pi} g_{\uparrow\downarrow} \left[\mathbf{m} \times \frac{\partial \mathbf{m}}{\partial t} \right], \tag{1.5}$$

where $g_{\uparrow\downarrow}$ is the spin-mixing conductance. As one can see the spin current $\mathbf{j}_{SP}$ is a direct current for uniform rotation of magnetization (with some constant frequency), consequently, such a dc spin current can transform to a dc electric current on the interface magnetic/metal due to the inverse spin Hall effect. Hence, both the spin pumping effect and the inverse spin Hall effect can be utilized for the development of ac signal detector on the base of magnetic/metal bilayers [14, 33, 49].

Magnetization dynamics in a SD is governed by the Landau–Lifshitz–Gilbert–Slonczewski equation [51]:

$$\frac{\partial \mathbf{M}}{\partial t} = \gamma [\mathbf{B}_{\text{eff}} \times \mathbf{M}] + \frac{\alpha}{M_s} \left[\mathbf{M} \times \frac{\partial \mathbf{M}}{\partial t} \right] + \frac{\sigma(\beta) I(t)}{M_s} [\mathbf{M} \times [\mathbf{M} \times \mathbf{p}]], \tag{1.6}$$

where $\mathbf{B}_{\text{eff}}$ is the effective magnetic field acting on the magnetization vector $\mathbf{M}$, α is the Gilbert damping constant.

The effective field $\mathbf{B}_{\text{eff}}$ in Eq. (1.6) can be written as a sum of several terms. Usually $\mathbf{B}_{\text{eff}}$ includes contributions from the external bias dc field $\mathbf{B}_{\text{dc}}$, the demagnetization (magneto-dipolar) field $\mathbf{B}_{\text{dip}}$, the anisotropy field $\mathbf{B}_{\text{a}}$, and the exchange field $\mathbf{B}_{\text{ex}}$.

In a general case the magnetization $\mathbf{M}$ in Eq. (1.6) is a function of time t and radius-vector $\mathbf{r}$. However, in nano-scale magnetic systems the so-called macrospin approximation is frequently used, and in accordance to it the magnetization vector can be considered as a function of time only. Thus, assuming $\mathbf{M} \equiv \mathbf{M}(t)$ and introducing the unit vector $\mathbf{m} \equiv \mathbf{m}(t) = \mathbf{M}(t)/M_s$, one can obtain a simplified equation for $\mathbf{m}$ from Eq. (1.6):

$$\frac{d\mathbf{m}}{dt} = \gamma [\mathbf{B}_{\text{eff}} \times \mathbf{m}] + \alpha \left[\mathbf{m} \times \frac{d\mathbf{m}}{dt} \right] + \sigma(\beta) I(t) [\mathbf{m} \times [\mathbf{m} \times \mathbf{p}]]. \tag{1.7}$$

1.3   Spin-Torque Diodes

There are two basic dynamical regimes of diode operation: traditional resonance regime and late discovered non-resonance regime.

In the traditional resonance regime of operation of a SD the spin-transfer torque excites a small-angle (and usually in-plane) magnetization precession about the equilibrium direction of the magnetization in the free layer of the magnetic nanostructure (the typical trajectory of the magnetization precession is shown by a red dashed curve in Fig. 1.1). The SD working in the traditional regime was studied experimentally [5, 10, 17, 31, 57] and theoretically [37, 42, 60] and has the following properties:

1. The diode works as a resonance-type signal detector with a resonance frequency that is close to the frequency of the ferromagnetic resonance $\omega_0 = 2\pi f_0$ of the free magnetic layer (see Fig. 1.2a);
2. The operation frequency bandwidth of the detector has an order of the ferromagnetic resonance linewidth Γ;

3. The diode's output dc voltage U_{dc} is proportional to the input ac power $P_{ac}\tilde{I}_{ac}^2$ (I_{ac} – ac current amplitude) and, consequently, the diode operates as a resonance-type quadratic detector (see inset in Fig. 1.2a):

$$U_{dc} = \varepsilon_{res} P_{ac} \frac{\Gamma^2}{\Gamma^2 + (\omega - \omega_0)^2},\qquad(1.8)$$

where $\omega = 2\pi f$ is the input signal angular frequency, ε_{res} is the resonance (at $\omega = \omega_0$) volt-watt sensitivity of the detector that strongly depends on the equilibrium angle

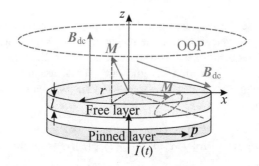

Fig. 1.1 The model of a SD: circular nano-pillar of radius r consists of the free layer of thickness l and the pinned magnetic layer. Under the action of ac current $I(t)$, the magnetization vector **M** (shown by a blue or red arrow) is precessing along small-angle in-plane (red dashed curve, resonance regime) or large-angle out-of-plane (blue dashed curve, OOP-regime) trajectory about the direction of the bias magnetic field $\mathbf{B}_{dc}$ (shown by a red or blue arrow), **p** is the unit vector in the direction of the magnetization of the pinned layer

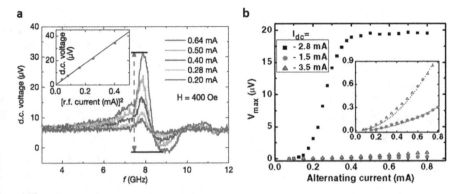

Fig. 1.2 (a) Typical frequency dependence of an output dc voltage of a SD operating in the resonance regime for different values of input ac current amplitude; the inset shows a typical dependence of the detector's output dc voltage on the input ac current amplitude (Reprinted by permission from Springer Nature: [57]). (b) Output dc voltage of a SD as a function of input ac current amplitude for a dc biased SD operating in the OOP-regime. (Reprinted figure with permission from American Physical Society: [6])

β_0 between the magnetization directions of the free and pinned layers. The resonance sensitivity of a traditional SD $\varepsilon_{res} = U_{dc}/P_{ac}$ is predicted to be approximately $\varepsilon_{res} \sim 10^4$ mV/mW for a passive (no dc bias) detector [60], while the best experimental value achieved to date is $\varepsilon_{res} = 630$ mV/mW for a conventional unbiased STMD [31] and $\varepsilon_{res} = 970$ mV/mW for a passive detector based on magnetic tunnel junction having a voltage-controlled interfacial perpendicular magnetic anisotropy of the free layer [10]. The resonance volt-watt sensitivity of a SD can be greatly enhanced by applying to the detector a dc bias current sufficiently large to compensate the natural damping in the free layer of a magnetic nanostructure. Recent experiments show that such dc-biased SDs may have the resonance volt-watt sensitivity of $\varepsilon_{res} \sim 1.2 \cdot 10^4$ mV/mW [31], $\varepsilon_{res} \sim 7.4 \cdot 10^4$ mV/mW [10] and even $\varepsilon_{res} \sim 2.2 \cdot 10^5$ mV/mW [5]. These values of the resonance volt-watt sensitivity ε_{res} of a SD are comparable to (passive detector) or greater than (dc-biased detector) the volt-watt sensitivity of a semiconductor Schottky diode [10].

Non-resonance Out-of-Plane Regime In contrast to the just described resonance regime of SD operation, now we consider the distinct SD operation regime based on the excitation of a current-driven large-angle out-of-plane (OOP) magnetization precession in the free layer of a SD (the corresponding trajectory of magnetization precession is shown by the blue dashed curve in Fig. 1.1). It was found theoretically that this regime of SD operation, the so-called OOP-regime, can be observed, when the SD is biased by the perpendicular dc magnetic field $\mathbf{B}_{dc} = \hat{\mathbf{z}}B_{dc}$, which is smaller than the saturation magnetic field of the FL ([41], see Fig. 1.1; $\hat{\mathbf{z}}$ is the unit vector of z-axis, which is perpendicular to the sample's plane). A SD performance in the OOP-regime qualitatively differs from its performance in the traditional resonance regime. In particular, excitation of magnetization oscillations in the FL does not have a resonance character in the OOP-regime and can be realized for a wide range of driving frequencies. Also, the output dc voltage of a SD working in the OOP regime is nearly independent of the input ac power, if the power exceeds a certain threshold value (Fig. 1.2b). The non-resonance regime of a SD operation was studied experimentally in [6, 11] and then explained theoretically in [37, 41, 42].

1.4   Spin Hall Detectors

There are several types of spin Hall effects described in details in [49] that can be used for the detection of spin currents excited by electromagnetic signals. Nowadays the signal detectors based on the inverse spin Hall effect become a standard measurement tool [14, 49]. In such detectors, input ac signal excites magnetization dynamics in the FL attached to a layer of metal having a strong spin-orbit interaction (like Pt, W, Au, Al, Mo, Pd, etc.) [49]. The excitation of magnetization oscillations can be usually caused due to the parametric ac pumping [33, 44] or the spin Hall effect [49]. In the first case, input ac signal of frequency f excites parametric magnons with the frequency $f/2$ in the nanostructure's FL. In the second case, input ac current subjected to a metal layer of the metal/magnetic bilayer structure is transformed to a transverse spin current due to the spin Hall effect that, in turn, excites the magnetization oscillations. The appearance of magnetization oscillations then can be detected by measuring a dc voltage generated in the metal layer due to the spin pumping and the inverse spin Hall effects [32, 33, 44, 49].

An example of measured inverse spin Hall dc voltage as a function of a bias dc magnetic field H_{dc} applied to a permalloy/Pt sample is shown in Fig. 1.3 (signal frequency is 4 GHz; [32]). The corresponding ac power dependence of the voltage is shown in the inset in Fig. 1.3. As one can see from Fig. 1.3 typical spin Hall detector operates as a resonance-type detector with resonance frequency f that is close to the

Fig. 1.3 Ac power dependence of the symmetric inverse spin Hall effect voltage contribution measured at 4 GHz in permalloy/Pt bilayer. The inset shows that the maximum of the inverse spin Hall effect signal is linear vs ac excitation power. (Reprinted figure with permission from American Physical Society: [32])

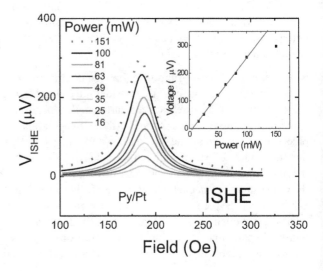

ferromagnetic resonance frequency f_0 of the used magnetic layer. And a rather large dc voltage is generated when the input signal frequency f coincides with the field-dependent ferromagnetic resonance frequency $f_0 \equiv f_0(H_{dc})$; by altering the bias dc magnetic field H_{dc} one can change the frequency operation range $[f_0 - \Gamma/2; f_0 + \Gamma/2]$ of the considered detector ($\Gamma \equiv \Gamma(H_{dc})$ is the ferromagnetic resonance linewidth). Also the detector works as a quadratic detector of low-power microwave signals, i.e. the measured inverse spin Hall dc voltage is proportional to the input ac power at least at small incident powers, when uniform magnetization modes are excited only [32].

1.5   Noise Properties of a Resonant-Type Spin-Torque Microwave Detector

It is obvious that the signal-to-noise ratio SNR and the minimum detectable power P_{min} of a resonant-type microwave detector based on a SD are limited by noise. To study the noise-handling properties of such a detector we developed a theory of the noise properties of a passive (no dc bias current) resonance-type spintronic detector based on the model proposed in [60] with additional terms describing influence of thermal fluctuations [37, 39, 42]. Note that experimental study of microwave properties of such detectors was done in [5, 10, 17, 31, 57]. In the theory we used the macrospin approximation, magnetization of the pinned layer is assumed to be truly "pinned", and we considered the case of a linear response of the magnetization to the ac current and thermal fluctuations. The detailed description of our study is presented in [42].

We took into account several possible sources of noise in a SD [42]:

(i)   Low-frequency Johnson-Nyquist noise is an additive source of noise and it is independent of the magnetization dynamics. Such low-frequency voltage fluctuations U_{JN} are associated with the equilibrium electrical resistance of the SD R_0.

(ii)   High-frequency Johnson-Nyquist noise, which is characterized by the voltage U_{IM}. This source of noise is non-additive and does not contribute to the output detected signal U_{dc} directly. This noise is caused by additional fluctuations of the magnetization direction in the FL, which, in turn, causes fluctuations of the electrical resistance of the detector due to the magnetoresistance effect (discussed above). After the mixing of the detected ac current with the resistance fluctuations a non-additive low-frequency noise voltage U_{IM} is generated.

(iii)   The last considered source of noise, the magnetic noise, is characterized by the low-frequency voltage U_{MN}. This type of noise is induced due to the existence of thermal fluctuations of the magnetization direction in the FL, which leads to the fluctuations of the electric resistance of the SD. After mixing with the input ac current this resistance fluctuations produce a low-frequency voltage U_{MN}.

Prokopenko and co-authors developed an analytical theory of the noise properties of a resonance-type SD [39, 42] and found that the root mean square fluctuations ΔU_{dc} of the detector's output voltage U_{dc} can be determined as

$$\Delta U_{dc} = U_{JN}\sqrt{1 + \frac{U_{dc}}{U_{IM}} + \frac{U_{dc}}{U_{MN}}} \approx \frac{P_{JN}}{\varepsilon_{res}}\sqrt{1 + \frac{P_{ac}}{P_{MN}}}. \tag{1.9}$$

In Eq. (1.9) it was taken into account that for detectors with typical experimental parameters $U_{IM} > > U_{JN}, U_{MN}$. Also the equation was written in terms of ac signal power $P_{ac} = U_{dc}/\varepsilon_{res}$ and noise powers $P_{JN} = U_{JN}/\varepsilon_{res}$, $P_{MN} = U_{MN}/\varepsilon_{res}$, where ε_{res} is the resonance volt-watt sensitivity (responsivity) of a SD.

It follows from Eq. (1.9) that the signal-to-noise ratio SNR can be estimated as

$$\text{SNR} = \frac{U_{dc}}{\Delta U_{dc}} = \frac{P_{ac}}{P_{JN}}\sqrt{\frac{P_{MN}}{P_{MN} + P_{ac}}}. \tag{1.10}$$

A simple analysis of Eqs. (1.9) and (1.10) demonstrates that there are two distinct regimes of operation of a resonance-type SD in the presence of thermal noise. The first regime corresponds to the case of relatively high frequencies and small powers of the input ac signal, when $P_{MN} \gg P_{ac}$. In this case the minimum detectable microwave power P_{min} is limited by the low-frequency Johnson-Nyquist noise, $P_{min} = P_{JN}$. Thus, the SNR is linearly increases with the input ac power, $\text{SNR} \cong P_{ac}/P_{JN}$. In the second regime that is observable at relatively low input signal frequencies, but large ac signal powers ($P_{ac} \gg P_{MN}$), the SNR of the SD nonlinearly depends on P_{ac}, $\text{SNR} \cong \sqrt{P_{ac}/P_{min}}$, and the minimum detectable power is limited by the magnetic noise, $P_{min} = P_{JN}^2/P_{MN}$.

The existence of two distinct regimes of SD operation is illustrated in Figs. 1.4 and 1.5. It can be seen that both curves (presented in Fig. 1.4 in logarithmic coordinates) demonstrate the clear change of slope from 1 to 1/2 in the region,

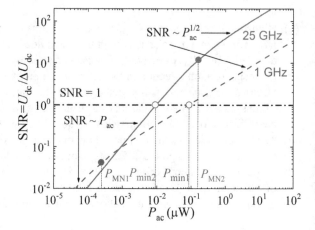

Fig. 1.4 Typical dependence of the signal-to-noise ratio of SD on the input microwave power calculated for two different frequencies of the input microwave signal: 1 GHz (dashed blue line) and 25 GHz (solid red line). P_{min} is the minimum detectable power of SD (at SNR = 1) and P_{MN} is the frequency-dependent characteristic power of magnetic noise. (Adapted from Ref. [39])

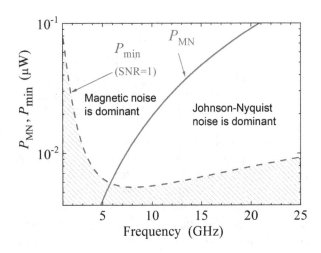

Fig. 1.5 The characteristic power of magnetic noise P_{MN} (solid red line) and minimum detectable power P_{min} of SD (dashed blue line) as functions of the input signal frequency. The blue dashed area corresponds to undetectable signals ($P_{ac} < P_{min}$). (Adapted from Ref. [39])

where the input ac power P_{ac} is close to the signal frequency-dependent character-istic power of the magnetic noise P_{MN}.

Figure 1.5 demonstrates the frequency dependence of the magnetic noise power P_{MN} and the minimum detectable microwave power P_{min}. The area below the curve $P_{MN}(f)$ is the area, where low-frequency Johnson-Nyquist noise is dominant, while in the area above $P_{MN}(f)$ the noise-handling properties of the SD mainly deteriorate due to existence of magnetic noise. The dashed area below $P_{min}(f)$ corresponds to the undetectable signals, and the minimum value of $P_{min}(f)$ is reached at the boundary between two discussed regimes, where $P_{JN} = P_{MN}(f)$.

We also analyzed the influence of temperature T on the performance of a passive SD and reveal that the detector's resonance volt-watt sensitivity, signal-to-noise ratio and minimum detectable ac power do not always improve with the decrease of temperature [40].

The performed theoretical analysis of noise properties of the SD operating in the resonance regime show that the detector has a performance comparable to that of conventional Schottky diodes [6]. Also it is found that the measurement of the SD signal-to-noise ratio in a wide range of microwave powers can be used to determine the spin-polarization efficiency η ($P_{MN} \sim 1/\eta$). Finally, the impedance mismatch between the SD and input microwave transmission line has been taken into account, and it is shown that the best minimum value of the minimum detectable microwave power P_{min} can be achieved for SD with some particular lateral sizes (radii) $\sim$ 150 nm [39, 42].

1.6 Power-Dependent Operation Regimes of SDs

In typical experiments SD operates as a frequency-selective quadratic microwave detector (quadratic regime, $U_{dc}\tilde{I}_{ac}^2$) [17, 57, 60] or as a threshold detector ($U_{dc} = \mathrm{const}$ if $I_{ac} \geq I_{th}$) [37, 38]. Other possible operation regimes of a resonance-type SD having non-quadratic dependence $U_{dc}(I_{ac})$ are unexplored yet and should be a subject of study.

In this section of the paper we report numerical macrospin study of the detector's operation regimes for a wide range of input ac signal powers. Our simulations are based on the solution of the Landau–Lifshitz–Gilbert–Slonczewski Eq. (1.7) and calculation of the detector's output dc voltage U_{dc} from Eq. (1.4).

The results of our simulations are shown as points in Fig. 1.6, while the solid red line is the approximation curve $U_{dc}\tilde{I}_{ac}^2$, and the approximation curve $U_{dc}{\sim}I_{ac}$ is shown by blue line. We reveal that there are two novel regimes of a SD operation characterized by a non-quadratic dependence of the detector's output dc voltage U_{dc} on the input microwave current I_{ac}, $U_{dc}I_{ac}^{\tilde{n}}$, where $n = 1$ (linear regime) or $1 < n < 2$ (transient regime). In Fig. 1.6 U_{cr} is the critical voltage value that corresponds to the boundary between the quadratic and transient regime, while I_{th} is the threshold current required for the excitation of magnetization dynamics in the SD [53].

It can be seen from Fig. 1.6 that quadratic regime of SD operation corresponds to a small-angle magnetization precession around some equilibrium magnetic state (see

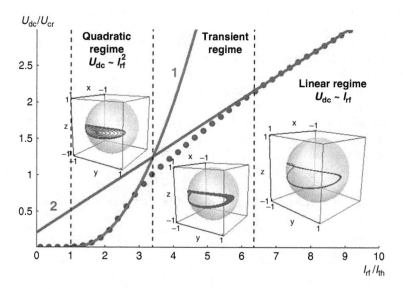

Fig. 1.6 Normalized output dc voltage of a SD as a function of its normalized input ac current. Points are the results of numerical simulations. Curves 1 and 2 are the quadratic (red) and linear (blue) approximations, respectively. Insets: qualitative magnetization dynamics trajectories realized for different operation regimes of a SD

magnetization precession trajectory on the unit sphere). Such precession can be considered as linear magnetization oscillations, i.e. the amplitude of precession is proportional to the input microwave power. However, with an increase of the input ac power, the nonlinearity of magnetization oscillations also increases, while the magnetization precession trajectory is transforming and approaching to some boundary cycle corresponded to the linear regime of detector operation.

We believe that the obtained results can be used for the calculation of dynamical range of powers of a SD and are important for the development and application of microwave devices based on such diodes.

1.7 Detector Based on an Array of SDs

A single resonance-type signal detector based on a SD cannot be used for precise microwave signal frequency measurement, because in accordance to Eq. (1.8) the measured detector's output dc voltage U_{dc} that is smaller than $\varepsilon_{res}P_{ac}$ corresponds to the two possible values of frequency $f_0 \pm \Delta f$, one that is smaller and other that is larger than the ferromagnetic resonance frequency f_0 [1]. This leads to the appearance of principal frequency measurement error that has an order of the ferromagnetic resonance linewidth (about several hundreds of megahertz in the microwave band).

To overcome the mentioned drawback of a single-diode measurement technique it can be used an advanced technique based on the application of an array of several SDs. In the simplest case of two independent SDs, when one have Eq. (1.8) for each uncoupled SDs, one can obtain the expression for the unknown input signal frequency and frequency measurement error [1]. A simple numerical calculations performed by using the derived expressions for two detectors with resonance frequencies of 3.0 GHz and 3.5 GHz demonstrate that (Fig. 1.7):

(a) The frequency measurement error Δf can be substantially smaller than the detector's linewidths $\Gamma_{1,\ 2}$, if SDs having different volt-watt sensitivities and linewidths, but relatively closely located resonance frequencies are used.
(b) The minimal values of frequency measurement error Δf can be achieved in the middle of the frequency range between detectors' resonance frequencies.
(c) The frequency measurement error decreases with the decrease of the FL thickness l, however detectors with very thin FLs (<1 nm) may have unsatisfactory noise properties.

Note that the use of an array with three different SDs instead of two also allows one to improve the precision of the frequency measurement [47].

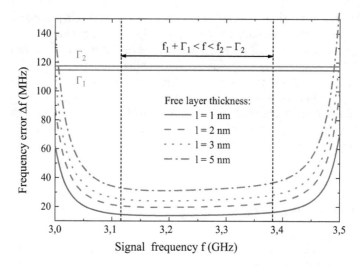

Fig. 1.7 The dependence of the frequency measurement error Δf on the frequency of input microwave signal f for different thicknesses of the free layer calculated for a SD with typical parameters. f_1 and f_2 are the frequencies and Γ_1 and Γ_2 are the linewidths of the first and the second SD, respectively. In the middle of the considered frequency range, the error of the input signal frequency measurement is several times smaller than the linewidths of both SDs. (Adapted from Ref. [1])

1.8 SD as a Threshold Spin-Torque Microwave Detector

In contrast to the well-known resonance regime of SD operation, here we consider a different regime of diode operation, based on the excitation of a large-angle out-of-plane (OOP) magnetization precession under the action of an input ac current $I(t)$ (see the blue dashed curve in Fig. 1.1). This regime is possible, when the SD is biased by the perpendicular magnetic field $\mathbf{B}_{dc} = \hat{\mathbf{z}}B_{dc}$, which is smaller than the saturation magnetic field of the FL [37, 41, 42].

We developed a theory of a SD operating in the OOP-regime [41], which was experimentally studied in [6]. It is based on the following assumptions: we consider a circular magnetic tunnel junction nano-pillar, where the magnetization of the FL is spatially-uniform, and the magnetization of the pinned layer is completely fixed and lie in the plane of the layer. For simplicity, we also neglect any in-plane anisotropy of the FL and assume that the junction is driven by a monochromatic input ac current only.

Using analytical and numerical calculations based on Eq. (1.7), one can obtain that the output voltage is generated in the OOP-regime if an input ac current (or power) exceeds a certain threshold. It can be also shown that the diode produces an output signal in a wide range of driving frequencies below some threshold frequency (Figs. 1.8 and 1.9; [41]). Finally, the output dc voltage of a SD in the OOP regime is almost constant, if the input ac power exceeds some power threshold,

Fig. 1.8 Typical dependence of the output dc voltage U_{DC} of a SD on frequency of input ac signal f in the OOP- (solid line and points) and resonance-regime (dashed line), respectively. Blue solid line and red dashed line are analytical dependencies. Points are the results of numerical simulations. Black crosses and green circles corresponded to the case when frequency is increased and decreased, respectively. (Adapted from Refs. [41, 42])

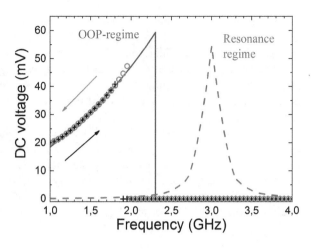

Fig. 1.9 Typical dependence of the output dc voltage U_{DC} of a SD on input ac current I_{RF} in the OOP- (solid line and points) and resonance-regime (dashed line), respectively. Blue solid line and red dashed line are analytical dependencies. Points are the results of numerical simulations. Black crosses and green circles corresponded to the case when the current is increased and decreased, respectively. (Adapted from Refs. [41, 42])

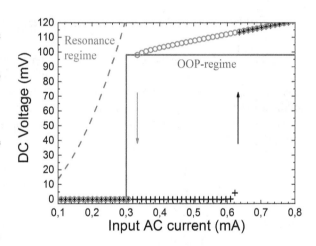

and it is zero for the powers below this threshold. In the first case of ac powers exceeding the threshold an output dc voltage of a SD can be written as [41].

$$U_{dc} \approx w I_{th}(\omega) R_{\perp},\qquad(1.11)$$

where

$$I_{th}(\omega) = 2\frac{\alpha}{v}\frac{\omega_M}{\sigma_{\perp}}\frac{\omega}{\omega_H - \omega}\qquad(1.12)$$

is the critical threshold current needed for excitation of the OOP-regime, w and v are small quantities dependent on the spin-polarization of current η, $R_{\perp}$ is the junction resistance in the perpendicular magnetic state ($\beta = \pi/2$), $\sigma_\perp$ is the angular-independent part of the Berger–Slonczewski's current-torque coefficient (in the case $\beta = \pi/2$), α is the Gilbert damping parameter, $\omega_M = \gamma\mu_0 M_s$ and $\omega_H = \gamma B_{dc}$ are the characteristic angular frequencies of a magnetic layer, and μ_0 is the vacuum permeability. The OOP-regime will be stable if the following approximate condition is fulfilled: $\omega < \omega_H$, where the threshold frequency is approximately equal to ω_H [37, 41, 42].

As one can see from Figs. 1.8 and 1.9, in the OOP-regime SD works as a non-resonance low-frequency threshold microwave detector. Its ac power dependence of output dc voltage has a step-like shape. Such a SD's behavior in the OOP-regime might also explain its extremely large diode volt-watt sensitivity ($\sim 10^5$ mV/mW) observed in experiments with thermally-activated "non-adiabatic stochastic resonance" [6], see also Fig. 1.2b.

1.9 Energy Harvesting of Ambient Microwave Signals

The SD in the OOP-regime can be used as a base element for novel energy harvesting devices designed for "Internet of Things" and RFID technologies, inasmuch as it has no resonance frequency, and, therefore, can absorb energy from all the low-frequency region ($\omega < \omega_H$) of the microwave spectrum.

The energy conversion rate ζ of such a harvester utilized a SD may be written as [41].

$$\zeta = \frac{P_{dc}}{P_{ac}} \approx \frac{1}{2}\left(\frac{I_{th}(\omega)}{I_{ac}}\right)^2 \left(\frac{w}{w_0}\right)^2 , \qquad (1.13)$$

where P_{dc} is the output dc power of a SD under the action of input microwave power P_{ac}, w and w_0 are small quantities dependent on the spin-polarization efficiency η [41, 42]. The maximum possible conversion rate $\zeta_{max} \approx 0.5w^2/w_0^2 \approx 3.5\%$ is reached in the case $I_{ac} = I_{th}(\omega)$.

Recent experiments performed for a SD based on a magnetic tunnel junction with perpendicular magnetic anisotropy of the FL operating in the OOP-regime has proved that such a harvester can operate in the absence of a bias dc magnetic field and can provide better performance than a semiconductor harvesting devices (Fig. 1.10) [11].

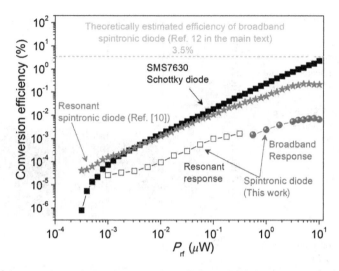

Fig. 1.10 A comparison of rf-to-dc conversion efficiency for Schottky diodes and spintronic diodes. The efficiencies in the Schottky diodes are calculated from the rectified dc voltages at 500 MHz. The efficiency in the resonant spintronic diode is calculated from the rectified dc voltages at the resonant frequency of 1.2 GHz. (Reprinted figure with permission from American Physical Society: [11])

1.10 Passive Demodulation of FM Digital Signals

Using the developed theory for a SD with perpendicular magnetic anisotropy of the FL operating in the OOP-regime, in [43] we proposed a theoretical model of a passive demodulator of low-power frequency-modulated signals based on such a SD. The device has several useful properties [27]:

(i) Rectified output voltage of a SD does not depend on the power of input signal and increases with the signal frequency, while this frequency is below some critical value;

(ii) The SD does not require an independent power supply and can operate by drawing the power from the input signal only;

(iii) The considered SD can operate in the absence of a bias dc magnetic field, thus it can be easily implemented in microelectronic systems, for example, CMOS;

(iv) The device can operate in the presence of input signals with rather small ac power (several nanowatts), thus it can be used in cases when Schottky diodes has relatively large input power threshold;

(v) The SD has an upper cutoff frequency and therefore it provides the transformation of input ac signals to a low-frequency output as a low-pass filter.

All statements mentioned above were proved by performing a numerical simulation for a SD with perpendicular magnetic anisotropy of the FL under the action of input frequency-modulated signal [27]. The demodulation of a signal encoded with frequency shift keying depicted in Fig. 1.11. A short digital data sequence is shown

Fig. 1.11 Demodulation of frequency shift keying (FSK) signal. (**a**) Input digital logic signal, where high levels indicate "1", and low state indicates "0". (**b**) The input logic signal encoded with FSK, where 60 MHz is "1", and 40 MHz is "0" Note that the amplitude increases from 6 to 8 μA. (**c**) Diode response to FSK input, 600 μV is "1" and 400 μV is "0". Note that the signal is dependent only on frequency and is independent of amplitude. (©2018 IEEE. Reprinted with permission, from [27])

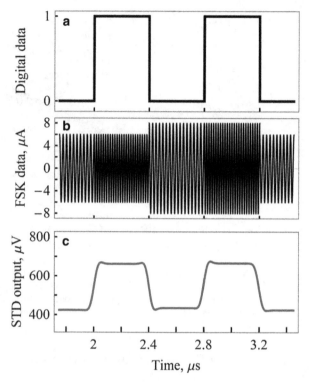

in Fig. 1.7 (a), where voltage varies between a high and a low state. Figure 1.11 (b) shows this data encoded in the frequency of a signal, "low" state is represented by input frequency $f_{in} = 40$ MHz and "high" state is represented by $f_{in} = 60$ MHz. The curve shown in Fig. 1.11(c) demonstrates that a SD acting on the modulated signal can faithfully reproduce a dc voltage representation of the input.

1.11 Spectrum Analyzer Based on a Current-Tunable Spin-Torque Nano-Oscillator

The use of a passive SD is not the only possible way to detect input microwave signals. In SD oscillations of input microwave current are mixing with oscillations of the SD resistance. But oscillations of input microwave current can be mixing with oscillations of a spin-torque nano-oscillator resistance as well. However, a spin-torque nano-oscillator allows one to easily tune frequency of the device resistance oscillations, thus providing means to scan certain frequency band and monitor the input signal spectrum in time. This case was analyzed in [28] where a spin-torque nano-oscillator was used instead of a SD.

Fig. 1.12 Block diagram of a spintronic spectrum analyzer. The magnetic tunnel junction based spin-torque nano-oscillator with tunneling magnetoresistance $r_{stno}(t)$, which is driven by ramped current I_{DC}, is multiplied by external microwave current $i_{ext}(t)$ to produce output voltage $v_{stno}(t)$. After passing through a low pass filter and a matched filter, a spectrum $v_{spec}(t)$ is produced. (Reprinted from Ref. [28], with the permission of AIP Publishing)

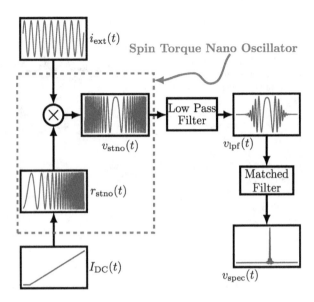

The block-diagram of the spectrum analyzer proposed is shown in Fig. 1.12 [28]. In this diagram the spin-torque nano-oscillator (outlined by a red dashed line) is driven by a ramped bias dc current $I_{DC}(t)$ that leads to the increase of a generated signal frequency $f(t)$ linearly with time. The tunneling magnetoresistance of the device, $r_{stno}(t)$, oscillates with the same frequency $f(t)$. An external microwave signal characterized by a current $i_{ext} = I_{ext} \cdot \cos(2\pi \cdot f_{ext} \cdot t)$ with magnitude I_{ext} and frequency f_{ext} is subjected to the oscillator. During operation of the oscillator $i_{ext}(t)$ and $r_{stno}(t)$ are combined (via Ohm's law) to produce output voltage $v_{stno}(t)$ which has a high frequency component $[f(t) + f_{ext}]$ and low frequency component $[f(t) - f_{ext}]$.

To get information about input signal spectrum the output voltage analyzed in two steps. First, an output voltage $v_{stno}(t)$ passed through a low pass filter having a cutoff frequency f_c leading to the elimination of the high frequency component of $v_{stno}(t)$. Second, previously obtained low-frequency voltage $v_{lpf}(t)$ ($v_{stno}(t)$ without the high frequency component) passed through a matched filter, which allows one to remove the dependence of the low-frequency voltage $v_{lpf}(t)$ on the phase difference between the mixed signals and improves the signal-to-noise ratio. The matched filter output, v_{spec}, provides the spectrum of $i_{ext}(t)$. One of important parameters of the device proposed is the resolution bandwidth or frequency resolution. It is a measure of the minimum separation required to distinguish two neighboring frequencies. The higher the scan rate ρ, the lower the resolution bandwidth. Response of the system proposed to a signal with the spectrum containing integer frequencies from 27 to 35 GHz with a random phase analyzed at a different scan-rate ρ and low-pass filter cutoff frequency f_c is presented in Fig. 1.13 [28].

The resolution bandwidth of the system analyzed in [28] is approximately equal to 220 MHz. Note that this resolution bandwidth is near the theoretical limit for scan

Fig. 1.13 Spectra produced by identical 10 nA signals at every integer from 27 to 35 GHz and a random phase. Responses from the following scan parameters: (a) $\rho = 1$ GHz/ns and $f_c = 4.5$ GHz, (b) $\rho = 1$ GHz/ns and $f_c = 6.4$ GHz, and (c) $\rho = 0.1$ GHz/ns and $f_c = 4.5$ GHz. (Reprinted from Ref. [28], with the permission of AIP Publishing)

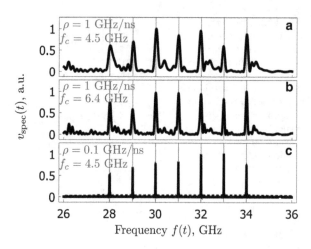

rates up to 5 GHz/ns. Also the sensitivity of a SD used for the spectrum analyzer has been found to be at the level of the Johnson-Nyquist thermal noise floor.

1.12 Antiferromagnetic Spin Current Rectifier

Despite the several advantages of SDs employing ferromagnets, typical operation frequencies of these detectors have an order of several GHz or several tens of GHz. It is almost impossible to make them operate at significantly higher frequencies, first of all in the terahertz gap. Hence, other types of devices and materials must be used. In particular, antiferromagnetic materials can be promising for TF applications, because due to a very strong internal exchange magnetic field the magnetic moments in antiferromagnets can precess with significantly larger frequencies than in ferromagnets, typically these frequencies lie in the range from 0.1 THz to 10.0 THz [3, 12, 13, 19, 22, 54].

A spin current rectification device working similar to a conventional ferromagnetic SD, but having substantially higher operation frequencies has been proposed recently [22]. In such a device input ac current acts on a Pt layer of a Pt/antiferromagnet bilayer structure and is transformed (due to the spin Hall effect) at the Pt/antiferromagnet interface to a perpendicular ac spin current flowing into the AFM layer. The ac spin current excites magnetization oscillations of magnetic sublattices in an antiferromagnet. These oscillations, in turn, caused the generation of a spin current flowing back to the interface antiferromagnet/Pt (the spin pumping effect), where the spin current is converted to an output dc current due to the inverse spin Hall effect. Thus the output dc current flowing in the Pt layer produces an output dc voltage across the Pt layer that can be measured.

In [22] it was shown that the output dc voltage U_{dc} of the considered rectifier can be estimated as

$$U_{dc} = J_{dc} L \theta_{SH} \rho \frac{\lambda_{Pt}}{d_{Pt}} \tanh \frac{d_{Pt}}{2\lambda_{Pt}}, \qquad (1.14)$$

where $J_{dc} \tilde{j}_{ac}^2$ is the output dc spin current density depending on the input electrical ac current density j_{ac}, L is the length of the Pt layer (distance between electrodes, where the output dc voltage is measured), θ_{SH} is the spin Hall angle (for details see [49]), ρ is the Pt resistivity, λ_{Pt} and d_{Pt} are the spin diffusion length in Pt and the thickness of the Pt layer, respectively. It follows from Eq. (1.14) that the antiferromagnetic rectifier works as a quadratic detector ($U_{dc} \tilde{j}_{ac}^2$). The calculations based on Eq. (1.14) performed in [22] have shown that the volt-watt sensitivity ε of the rectifier reaches its maximum value ε_{res} at the frequencies close to the antiferromagnetic resonance frequency of the used antiferromagnet, so the rectifier may work as a quasi-resonant device. Also it was found that ε nonlinearly depends on the sample's size L, and it reaches its maximum at some frequency-dependent value of L [22].

Thus, in [22] it was shown that an antiferromagnet with biaxial anisotropy (e.g., NiO) can be used as an active element for the ac-dc conversion of spin currents providing the sensitivity of such a SD is in the range of $\varepsilon_{res} \sim 10^2 - 10^3$ mV/mW, which is comparable with the sensitivity of modern electrical Schottky diodes. Also it was demonstrated that both the sign and magnitude of the rectified dc current are determined by the mutual arrangement of the crystallographic axes of the antiferromagnet and the direction of polarization of the input ac spin current. Note that the presence of the bi-axial anisotropy is crucial for the spin rectification, since the effect is caused by the angular momentum exchange between the spin sub-system and the crystal lattice of the antiferromagnet, and is absent in uniaxial antiferromagnetic materials [22].

1.13   Detector and Spectrum Analyzer Based on a Current-Driven Antiferromagnetic Tunnel Junction

In this section of the paper we theoretically consider an efficient method to perform spectrum analysis on low power signals in the frequency range from 0.1 to 10 THz [2]. The method uses a nano-scale antiferromagnetic tunnel junction (Fig. 1.14) having a tunneling anisotropic magnetoresistance $R(t)$ that is oscillating at a frequency $f(t)$ depending on the magnitude of the drive current I_{drive}:

$$R(t) = R_0 + \Delta R \cos\left(2\pi \int f(t) dt\right), \qquad (1.15)$$

where the frequency $f(t)$ increases linearly with time t, when the drive current I_{drive} increases, $f(t) = f_0 + \rho t$, f_0 is the initial frequency and ρ is the rate of the frequency

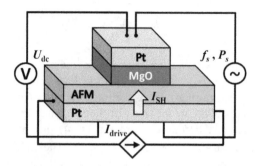

Fig. 1.14 Schematic view of an antiferromagnetic tunnel junction. I_{drive} is the driving current; I_{SH} is the spin current created by the drive current due to the spin Hall effect; f_s and P_s denote frequency and power of the input signal, respectively, and U_{dc} is the output dc voltage generated by mixing of the junction resistance oscillations with an input ac current

change. When these current-driven resistance oscillations $R(t)$ are mixed with an input signal current $I_s \cos(2\pi f_s t + \phi)$, an output voltage

$$U_{out} = \frac{1}{2} I_s \Delta R \cos(\theta(f_s, t) - \phi), \qquad (1.16)$$

is generated; here $\theta(f_s, t) = 2\pi(f_0 - f_s)t + \pi\rho t^2$. This signal then passes through a low-pass filter and one can obtain a low-frequency voltage. Finally, a matching filter technique is used to remove the dependence of this low-frequency output signal on the initial phase ϕ [2].

To calculate an output dc voltage of the device that follows from Eq. (1.16) we use a simple electrical model of an antiferromagnetic tunnel junction [55]. We consider this junction as an oscillating resistance $R(t)$ connected in parallel to the junction capacitance with a capacity C. We assume that the junction circuit with frequency-dependent impedance is connected to an external circuit with 50 Ω impedance via an ideal bias tee and calculate an input current magnitude I_s for known input signal power with an account of the impedance mismatch effect. Then assuming that the signal frequency f is fixed we calculate $U_{out}(f)$. We also found that the junction performance strongly depends on the thickness of MgO dielectric layer d and the junction's cross-sectional area S, because $\Delta R \sim R_0 \sim \exp(\kappa d)/S$ and $C \sim S/d$; in [2] optimal frequency-dependent values of d and S providing the maximum value of $U_{out}(f)$ are determined. Finally, by using the spectrum measurement technique proposed in [28] based on the application of a low pass filter, and a matched filter, we have shown a possibility to perform the spectrum analysis of low-power ac signals in a relatively wide frequency range [2].

Our theoretical and numerical estimations show that the proposed spectrum analyzer can perform scanning over a 0.25 THz bandwidth in just 25 ns on signals having power at a level of the Johnson-Nyquist thermal noise floor (Fig. 1.15) [2]. We also demonstrate that using the typical parameters from [34] an output dc voltage of the device has been estimated to be rather low (about 0.1 μV). However,

Fig. 1.15 Frequency dependence of the output dc voltage (blue line) and the Johnson-Nyquist thermal noise floor rms voltage (red line) for an antiferromagnetic tunnel junction (ATJ) with typical parameters under the action of input ac power of 1 μW. The dashed area corresponds to ac signals with the signal-to-noise ratio below 1. (Adapted from [2])

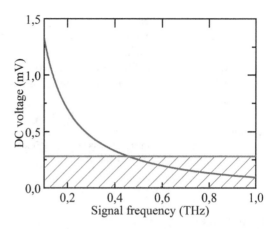

by choosing the junction with a thin MgO tunneling barrier ($d = 1$ nm) and a relatively small cross-sectional area ($S = 300 \times 300$ nm^2) one can obtain the output dc voltage of about 1 mV in the frequency range 0.1–1 THz for incident ac power of 1 μW that should be sufficient for some practical applications. It should be noted that in contrast to the spin current rectifier [22], the proposed device does not require any special conditions for the antiferromagnetic layer.

Finally, we would like to note although the performance of the considered TF antiferromagnetic rectifier and spectrum analyzer is not very good at the moment, there are a lot of possibilities to improve their performance (by applying an additional bias dc current / magnetic field to an antiferromagnet, use antiferromagnetic materials with additional anisotropy, better material parameters, utilize a more efficient measurement technique, etc.). Thus, we believe that the proposed antiferromagnetic nanoscale devices are promising for the signal detection and processing in the TF gap and could be widely used after a proper optimization of their technical characteristics.

1.14   Conclusions

In conclusion, we have considered basic phenomena used in different types of spintronic diodes based on ferromagnets and/or antiferromagnets operating in the microwave and terahertz band. We described the principles of operation of spin-torque and spin Hall detectors and analyzed their typical properties in different operation regimes. It has been demonstrated that such diodes are able to detect ac signals with volt-watt sensitivity that exceeds that of semiconductor Schottky diodes and can be used for fast and precise spectrum analysis. We have also considered a precise frequency measurement technique with two uncoupled SDs, and a novel algorithm of signal detection based on the application of a dc biased spin-torque nano-oscillator. In addition to that we have demonstrated a possibility of demodulation of frequency-modulated signals and microwave energy harvesting in ferromagnetic SDs. Finally, we have theoretically proposed a novel signal detection technique based on the application of a spin Hall rectifier utilizing a bi-axial antiferromagnet (e.g., NiO) and a prospective terahertz-frequency spectrum analysis technique based on the application of a current-driven antiferromagnetic tunnel junction.

Acknowledgments This work was supported in part by the grants Nos. 18BF052–01 M and 19BF052–01 from the Ministry of Education and Science of Ukraine and grant No. 1F from the National Academy of Sciences of Ukraine.

References

1. Artemchuk PY, Prokopenko OV (2017) Microwave signal frequency determination using multiple spin-torque microwave detectors. In: Proceedings of the IEEE First Ukraine Conference on Electrical and Computer Engineering (UKRCON), Kyiv, pp 628–633
2. Artemchuk PY, Sulymenko OR, Louis S et al (2020) Terahertz frequency spectrum analysis with a nanoscale antiferromagnetic tunnel junction. J Appl Phys 127:063905
3. Baltz V, Manchon A, Tsoi M et al (2018) Antiferromagnetic spintronics. Rev Mod Phys 90:015005
4. Berger L (1996) Emission of spin waves by a magnetic multilayer traversed by a current. Phys Rev B 54:9353
5. Carpentieri M, Zeng Z, Finocchio G (2016) Giant spin-torque diode sensitivity in the absence of bias magnetic field. In: Abstracts of Baltic Spin Conference, Jurmala, 9–13 August 2016
6. Cheng X, Boone CT, Zhu J et al (2010) Nonadiabatic stochastic resonance of a nanomagnet excited by spin torque. Phys Rev Lett 105:047202
7. Cheng R, Xiao D, Brataas A (2016) Terahertz Antiferromagnetic Spin Hall Nano-Oscillator. Phys Rev Lett 116:207603
8. Demidov VE, Urazhdin S, Ulrichs H et al (2012) Magnetic nano-oscillator driven by pure spin current. Nature Mater 11:1028
9. Dhillon SS, Vitiello MS, Linfield EH et al (2017) The 2017 terahertz science and technology roadmap. J Phys D: Appl Phys 50:043001
10. Fang B, Carpentieri M, Hao X et al (2016) Giant spin-torque diode sensitivity in the absence of bias magnetic field. Nature Commun 7:11259
11. Fang B, Carpentieri M, Louis S et al (2019) Experimental demonstration of spintronic broadband microwave detectors and their capability for powering nanodevices. Phys Rev Appl 11:014022
12. Gomonay HV, Loktev VM (2010) Spin transfer and current-induced switching in antiferromagnets. Phys Rev B 81:144427
13. Gomonay HV, Loktev VM (2014) Spintronics of antiferromagnetic systems (Review Article). Low Temp Phys 40:17
14. Hahn C, de Loubens G, Viret M et al (2013) Detection of microwave spin pumping using the inverse spin hall effect. Phys Rev Lett 111:217204
15. Houssameddine D, Ebels U, Delaet B et al (2007) Spin-torque oscillator using a perpendicular polarizer and a planar free layer. Nat Mater 6:447
16. Hübers HW (2010) Terahertz technology: Towards THz integrated photonics. Nat Photon 4:503
17. Ishibashi S, Seki T, Nozaki T et al (2010) Large diode sensitivity of CoFeB/MgO/CoFeB magnetic tunnel junctions. Appl Phys Express 3:073001

18. Ivanyuta OM, Prokopenko OV, Raksha VM et al (2005) Microwave detection using Josephson junction arrays integrated in a resonator. Physica Status Solidi C 2:1688
19. Jungwirth T, Marti X, Wadley P et al (2016) Antiferromagnetic spintronics. Nat Nanotech 11:231
20. Kaka S, Pufall MR, Rippard WH et al (2005) Mutual phase-locking of microwave spin torque nano-oscillators. Nature (London) 437:389
21. Katine JA, Albert FJ, Buhrman RA et al (2000) Current-driven magnetization reversal and spin-wave excitations in Co/Cu/Co pillars. Phys Rev Lett 84:3149
22. Khymyn R, Tiberkevich V, Slavin A (2017) Antiferromagnetic spin current rectifier. AIP Adv 7:055931
23. Kiselev SI, Sankey JC, Krivorotov IN et al (2003) Microwave oscillations of a nanomagnet driven by a spin-polarized current. Nature 425:380
24. Kleiner R (2007) Filling the Terahertz Gap. Science 318:1254
25. Krivorotov IN, Emley NC, Sankey JC et al (2005) Time-domain measurements of nanomagnet dynamics driven by spin-transfer torques. Science 307:228
26. Lee KJ, Deac A, Redon O et al (2004) Excitations of incoherent spin-waves due to spin-transfer torque. Nat Mater 3:877
27. Louis S, Tiberkevich V S, Li J et al (2018) Spin torque diode with perpendicular anisotropy used for passive demodulation of FM digital signals. In: Proceedings of the IEEE International Magnetics Conference (INTERMAG), Singapore
28. Louis S, Sulymenko O, Tiberkevich V et al (2018) Ultra-fast wide band spectrum analyzer based on a rapidly tuned spin-torque nano-oscillator. Appl Phys Lett 113:112401
29. Malyshev V, Melkov G, Prokopenko O (2020) Microwave devices based on superconducting surface electromagnetic wave resonator (Review article). Low Temp Phys 46:348
30. Mancoff FB, Rizzo ND, Engel BN et al (2005) Phase-locking in double-point-contact spin-transfer devices. Nature 437:393
31. Miwa S, Ishibashi S, Tomita H et al (2014) Highly sensitive nanoscale spin-torque diode. Nature Mater 13:50
32. Mosendz O, Vlaminck V, Pearson JE et al (2010) Detection and quantification of inverse spin Hall effect from spin pumping in permalloy/normal metal bilayers. Phys Rev B 82:214403
33. Noack TB, Vasyuchka VI, Bozhko DA et al (2019) Enhancement of the spin pumping effect by magnon confluence process in YIG/Pt bilayers. Physica Status Solidi B 1900121
34. Park BG, Wunderlich J, Marti X et al (2011) A spin-valve-like magnetoresistance of an antiferromagnet-based tunnel junction. Nat Mater 10:347
35. Pozar DM (1998) Microwave engineering. Wiley, New York
36. Prokopenko OV (2015) Microwave signal sources based on spin-torque nano-oscillators. Ukr J Phys 60:104
37. Prokopenko OV, Slavin AN (2015) Microwave detectors based on the spin-torque diode effect. Low Temp Phys 41:353
38. Prokopenko OV, Bankowski E, Meitzler T et al (2011) Spin-torque nano-oscillator as a microwave signal source. IEEE Magn Lett 2:3000104
39. Prokopenko OV, Melkov G, Bankowski E et al (2011) Noise properties of a resonance-type spin-torque microwave detector. Appl Phys Lett 99:032507
40. Prokopenko OV, Bankowski E, Meitzler T et al (2012) Influence of temperature on the performance of a spin-torque microwave detector. IEEE Trans Magn 48:3807
41. Prokopenko OV, Krivorotov IV, Bankowski E et al (2012) Spin-torque microwave detectors with out-of-plane precessing magnetic moment. J Appl Phys 111:123904
42. Prokopenko OV, Krivorotov IN, Meitzler TJ et al (2013) Spin Torque Microwave Detectors. In: Demokritov SO, Slavin AN (eds) Magnonics: from fundamentals to applications, Topics in applied physics, vol 125. Springer, Berlin/Heidelberg, pp 143–161
43. Prokopenko OV, Tyberkevych V, Slavin A (2017) Microwave energy harvester based on a spin-torque diode with perpendicular magnetic anisotropy of the free layer. In: Proceedings of MMM conference, Pittsburg, November 2017

44. Prokopenko OV, Bozhko DA, Tyberkevych VS et al (2019) Recent trends in microwave magnetism and superconductivity. Ukr J Phys 64:888
45. Ramo S, Whinnery JR, van Duzer T (1994) Fields and waves in communication electronics. Wiley, New York
46. Ruotolo A, Cros V, Georges B et al (2009) Phase-locking of magnetic vortices mediated by antivortices. Nat Nanotechnol 4:528
47. Shanidze RG, Prokopenko OV (2018) Determination of microwave signal frequency in a system of three uncoupled spin-torque microwave detectors. In: Proceedings of the International Young Scientists' Conference on Applied Physics, Kiev, 2018
48. Shaternik V, Belogolovskii M, Prikhna T et al (2012) Universal character of tunnel conductivity of metal-insulator-metal heterostructures with nanosized oxide barriers. Physics Procedia 36:94
49. Sinova J, Valenzuela SO, Wunderlich J et al (2015) Spin Hall effects. Rev Mod Phys 87:1213
50. Sirtori C (2002) Applied physics: bridge for the terahertz gap. Nature 417:132
51. Slavin A, Tiberkevich V (2009) Nonlinear auto-oscillator theory of microwave generation by spin polarized current. IEEE Trans Magn 45:1875
52. Slonczewski JC (1996) Current-driven excitation of magnetic multilayers. J Magn Magn Mat 159:L1
53. Sorokin S, Prokopenko OV (2016) Linear and nonlinear operation regimes of a spin-torque microwave detector. In: Abstracts of the Baltic Spin conference, Jurmala, 9–13 August 2016
54. Sulymenko OR, Prokopenko OV, Tyberkevych VS et al (2017) Terahertz-frequency spin hall auto-oscillator based on a canted antiferromagnet. Phys Rev Appl 8:064007
55. Sulymenko OR, Prokopenko OV, Tyberkevych VS et al (2018) Terahertz-Frequency Signal Source Based on an Antiferromagnetic Tunnel Junction. IEEE Magn Lett 9:3104605
56. Tonouchi M (2007) Cutting-edge terahertz technology. Nat Photon 1:97
57. Tulapurkar AA, Suzuki Y, Fukushima A et al (2005) Spin-torque diode effect in magnetic tunnel junctions. Nature 438:339
58. Urazhdin S, Birge NO, Pratt WT et al (2003) Current-driven magnetic excitations in permalloy-based multilayer nanopillars. Phys Rev Lett 91:146803
59. Wang C, Cui YT, Sun JZ et al (2009) Bias and angular dependence of spin-transfer torque in magnetic tunnel junctions. Phys Rev B 79:224416
60. Wang C, Cui YT, Sun JZ (2009) Sensitivity of spin-torque diodes for frequency-tunable resonant microwave detection. J Appl Phys 106:053905

Chapter 2
Microwave Characterization of $Y_3Fe_5O_{12}$ Ferrite Under a dc-Magnetic Field

Erene Varouti, Euthimios Manios, Ilias Tsiachristos, Antonis Alexandridis, and Michael Pissas

Abstract The scattering parameters (S_{11}, S_{21}) for a rectangular waveguide, loaded with dielectric gyrotropic $Y_3Fe_5O_{12}$ ferrite were measured, targeting to estimate the electric permittivity and the parameters of the permeability tensor. The measurements were collected with a vector network analyzer equipped with a calibrated rectangular wave guide. For non-magnetic dielectric materials the electric permittivity and the corresponding $\tan\delta$ were estimated by "trial and error" method, comparing the exact calculated scattering parameters with the experimental ones, supposing a linear or quadratic frequency dependence of $\varepsilon(f)$. The theoretical relations describing the scattering parameters for a non-magnetic dielectric were calculated based only on the TE_{10} mode, for a rectangular wave guide. In the case of a ferromagnetic dielectric the parameter describing the losses of the magnetic permeability tensor, ΔH, under a dc-magnetic field, was estimated by trial and error method, comparing the experimental and theoretically calculated scattering parameters. The theoretical scattering parameters were calculated with a Finite Element electromagnetic simulation software. The dc-magnetic field was applied parallel to the short dimension of the waveguide and its amplitude was higher than the field needed to saturate the magnetization.

Keywords Microwave ferrites · Ferromagnetic resonance · Permeability tensor · Y3Fe5O12 · Scattering parameters · Vector network analyser

E. Varouti · E. Manios · I. Tsiachristos · M. Pissas (✉)
Institute of Nanoscience and Nanotechnology, NCSR "Demokritos", Athens, Greece
e-mail: m.pissas@inn.demokritos.gr

A. Alexandridis
Institute of Informatics and Telecommunications, NCSR "Demokritos", Athens, Greece

© Springer Nature B.V. 2020
A. Kaidatzis et al. (eds.), *Modern Magnetic and Spintronic Materials*, NATO
Science for Peace and Security Series B: Physics and Biophysics,
https://doi.org/10.1007/978-94-024-2034-0_2

2.1 Introduction

The propagation of electromagnetic waves in magnetically polarized dielectric media has been thoroughly studied in the past from theoretical point of view [3, 4, 6–10, 13, 17–21, 24].

A large number of devices, operating in the microwave regime of the electromagnetic spectrum, have been constructed using ferromagnetic materials. As typical examples we can mention the Y-circulators, Faraday circulator, isolators and phase shifters [see 6, 13, 17 for more details]. Recently, magnetically polarized materials have been used [1, 31] as part of the substrate to construct reconfigurable microstrip patch antennas. Moreover, ferrimagnetic insulated materials have been used in four port double Y-shaped ultra-wideband magneto-photonic crystals circulators for 5G communication systems [29].

The novel material characteristic that acquires a magnetically polarized, electrically insulating, ferromagnetic material, is that the magnetic permeability parameter μ becomes a complex antisymmetric tensor. When the dc-magnetization is point out along the z-axis the permeability tensor has the form [11–13, 16].

$$\mu = \begin{pmatrix} \mu & j\kappa & 0 \\ -j\kappa & \mu_0 & 0 \\ 0 & 0 & \mu_0 \end{pmatrix}$$

where, $\mu = \mu' - j\mu''$, and $\kappa = \kappa' - j\kappa''$. The physical origin of the tensor character of the permeability is related to the resonance motion of the ac-magnetization.

The purpose of this chapter is to present an experimental method for the estimation of the dielectric permittivity and the ferromagnetic resonance parameters, for bulk ferromagnetic material fully loading a rectangular wave guide.

2.2 Theoretical Background

The equation of motion of the magnetization in the presence of a dc magnetic field is given by [12, 13, 22, 23].

$$\frac{d\mathbf{M}}{dt} = -\gamma\mu_0\mathbf{M} \times \mathbf{H}_{eff} + \frac{\lambda\gamma\mu_0}{M_s}\mathbf{M} \times \left(\mathbf{M} \times \mathbf{H}_{eff}\right) \qquad (2.1)$$

(Landau and Lifshitz form) or equivalent

$$\frac{d\mathbf{M}}{dt} = -\gamma\mu_0\mathbf{M} \times \mathbf{H}_{eff} + \frac{\alpha}{M_s}\mathbf{M} \times \left(\mathbf{M} \times \frac{d\mathbf{M}}{dt}\right) \qquad (2.2)$$

(Gilbert form), where **M** is the magnetization, $\mathbf{H}_{\mathbf{eff}}$ is the effective magnetic field, γ is the gyromagnetic ratio of the electron (e.g. $\gamma = f/H_0 = g_{eff}\mu_0 \mid q_e \mid / (4\pi m_e) = 2.8$ MHz/Oe, μ_0 is the permeability of the free space. The λ and α are small parameters describing the losses. The parameter α is connected with the width of the absorption line, measured in ferromagnetic resonance experiments.

If the ferromagnetic material is magnetized by a dc magnetic field much larger than the ac-magnetic field, then under these small signal conditions

$$
\begin{aligned}
&\mid \mathbf{M}_{dc} \mid \ll \mid \mathbf{M}_{dc} \mid, \\
&\mid \mathbf{H}_{ac} \mid \ll \mid \mathbf{H}_{dc} \mid, \\
&\mathbf{M}_{ac} \times \mathbf{H}_{ac} \sim 0, \\
&\mathbf{M}_{dc} \| \mathbf{H}_{dc}
\end{aligned}
\tag{2.3}
$$

the Eqs. 2.1 and 2.2 are equivalent, by assuming that the ferromagnetic medium can be described [6, 12, 17], by a non-symmetrical permeability tensor (Polder tensor [12, 16]). For $\mathbf{H}_{dc} \| $ y-axis this tensor has the form

$$
[\mu] = \begin{pmatrix} \mu & 0 & j\kappa \\ 0 & \mu_0 & 0 \\ -j\kappa & 0 & \mu \end{pmatrix}
\tag{2.4}
$$

where, $\mu = \mu' - j\mu''$, $\kappa = \kappa' - j\kappa''$, and

$$
\mu' = \mu_0\left[1 + \frac{\omega_0\omega_m\left(\omega_0^2 - \omega^2\right) + \omega_0\omega_m\omega^2\alpha^2}{\left[\omega_0^2 - \omega^2(1 + \alpha^2)\right]^2 + 4\omega_0^2\omega^2\alpha^2}\right]
$$

$$
\mu'' = \mu_0 \frac{\alpha\omega\omega_m\left[\omega_0^2 + \omega^2(1 + \alpha^2)\right]}{\left[\omega_0^2 - \omega^2(1 + \alpha^2)\right]^2 + 4\omega_0^2\omega^2\alpha^2}
$$

$$
\kappa' = \mu_0\omega\omega_m\left[\frac{\left(\omega_0^2 - \omega^2(1 + \alpha^2)\right)}{\left[\omega_0^2 - \omega^2(1 + \alpha^2)\right]^2 + 4\omega_0^2\omega^2\alpha^2}\right. \kappa'' = \mu_0 \frac{2\omega_0\omega_m\omega^2\alpha}{\left[\omega_0^2 - \omega^2(1 + \alpha^2)\right]^2 + 4\omega_0^2\omega^2\alpha^2},
$$

$$
\tag{2.5}
$$

In previous relations $\omega_0 = \gamma\mu_0 H_0$ is the Larmor precession frequency, $\omega_m = \gamma\mu_0 M_s$, M_s the saturation magnetization, and $\alpha = \mu_0\gamma\Delta H/4\pi f_{fmr}$ is a phenomenological parameter describing the losses. f_{fmr} denotes the frequency used to measure the ΔH.

The propagation of the electromagnetic wave inside a fully loaded with ferromagnetic material a wave guide is described from the source free Maxwell equations

$$\nabla \cdot \mathbf{D} = 0$$
$$\nabla \cdot \mathbf{B} = 0$$
$$\nabla \times \mathbf{E} = -\frac{1}{c}\frac{\partial \mathbf{B}}{\partial t} \tag{2.6}$$
$$\nabla \times \mathbf{H} = \frac{\partial \mathbf{D}}{\partial t}$$

Supposing that $\mathbf{D} = \varepsilon\mathbf{E}$ and $\mathbf{B} = [\mu]\mathbf{H}$, for harmonically temporally varying fields (e.g. $\mathbf{E}(\mathbf{x}, t) = \mathrm{Re}\left[\widetilde{\mathbf{E}}(\mathbf{x})\exp(-i\omega t)\right]$, where $\widetilde{\mathbf{E}}(\mathbf{x})$ the phasor of the electric field) the Maxwell equations becomes equivalent to

$$\widetilde{\mathbf{B}} = -i\frac{c}{\omega}\nabla \times \widetilde{\mathbf{E}}$$
$$\nabla \times \left[[\mu]^{-1}\nabla \times \widetilde{\mathbf{E}}\right] - \frac{\omega\varepsilon}{c}\widetilde{\mathbf{E}} = 0 \tag{2.7}$$

$[\mu]^{-1}$ is the inverse of the magnetic permeability tensor.

2.3 Sample Preparation and Characterization

The samples were prepared by standard solid state reaction. Stoichiometric amounts of Fe_2O_3 and Y_2O_3 powders were thoroughly mixed in agate mortar [28]. The mixed powders are pressed in disk shape (see Fig. 2.1) and initially heated at 1350 °C to 1450 °C. The process of grinding palletization and high temperature reaction was repeated several times, in order to achieve the maximum possible homogenization of the stoichiometry. The crystallinity and the quality of the produced $Y_3Fe_5O_{12}$

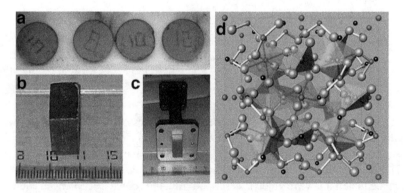

Fig. 2.1 (a) Cylindrical pellets of $Y_3Fe_5O_{12}$ compound after the final reaction step. (b) rectangular pellet after the sintering. (c) polished rectangular pellet inside the wave guide. (d) Unit cell of $Y_3Fe_5O_{12}$ compound. Iron ions Fe^{3+} occupy two crystallographic sites tetrahedrally and octahedrally coordinated with oxygen ions

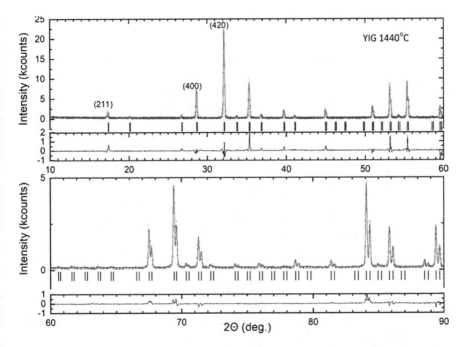

Fig. 2.2 Rietveld plot of the $Y_3Fe_5O_{12}$ compound (upper panel 2Θ range $10–60°$ and lower panel $60–90°$). The powder X-ray diffraction data were refined using the Rietveld method by employing the Fullprof suite of programs

compound were checked with x-ray diffraction data, Mossbauer spectra and magnetic measurements. Using the obtained optimum $Y_3Fe_5O_{12}$ powders, we prepared blocks with rectangular parallelepiped shape (see Fig. 2.1), using a die. Subsequently the blocks were sintered at 1400 °C for 48 h. The phase purity and crystallinity of the samples is checked by powder x-ray diffraction and magnetization measurements. X-ray powder diffraction (XRD) measurements were performed in Bragg-Brentano geometry using CuKαradiation and a graphite crystal monochromator in a Siemens D500 diffractometer, where $\theta - 2\theta$ scans between $10°$ and$120°$, in steps of$0.03°$, were measured with very-high statistics at ambient conditions. Fig. 2.2 shows the Rietveld plot of the sample used in wave guide measurements. The x-ray data were refined using the Fullprof suite of programs [5], adopting the crystal structure of $Y_3Fe_5O_{12}$ which is based on the $Ia\bar{3}d$ space group. The results of the refinements showed single phase material, with good crystallinity. The length of the cubic unit cell parameter was estimated to be $a = 1.23759$ nm in perfect agreement with the literature data [28 and references therein].

The samples are also characterized by dc and ac magnetic measurement. Figures 2.3 and 2.4 show the variation of the dc-magnetic moment per gram as a function of the dc-magnetic field, measured at room and liquid helium temperatures. The magnetization loops revealed a small coercive field about $H_c = 14 \pm 5$ Oe ~14 Oe (see Fig. 2.4b).

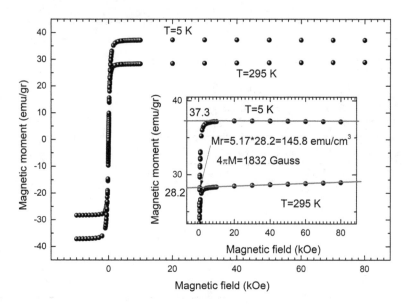

Fig. 2.3 Magnetic field variation of the magnetic moment per g of $Y_3Fe_5O_{12}$, measured at ambient temperature (T = 295 K) and at liquid helium temperatures (T = 5 K). The main panel shows the results of the measurements in full magnetic field scale. The inset depicts a detail of the data at saturation level of the magnetization. The saturation magnetization was estimated 28.2 emu/gr, and 37.3 emu/gr at 295 K and 5 K, respectively

A scanning electron microscope (SEM), Jeol JSM-6610, was employed for inspection of the microstructure in both types of samples. Figure 2.5 displays the scanning electron microscopy (SEM) picture displaying the surface morphology of the sintering samples. The surface morphology reveals polygonal grains of various sizes, with small porosity. The main grain size is about 10 μm.

The ΔH parameter was estimated from ferromagnetic resonance spectra measured using a small piece of the $Y_3Fe_5O_{12}$ compound by employing Bruker, ER 200D-SRC instrument. Figure 2.6, shows the derivative and the integrated absorption of the ferromagnetic resonance spectrum of a small bulk piece extracted from a large $Y_3Fe_5O_{12}$ block.

2.4   Measurements of PMMA and $Y_3Fe_5O_{12}$ in Zero Magnetic Field

Usually, in thin samples ($d \ll \lambda$, where d is the length of the sample parallel to the waveguide and λ the wave length) the wide band effective electric permittivity and magnetic permeability are measured with the Nicolson-Ross method [14]. Here we present the measurements of the scattering parameters of a rectangular wave guide completely loaded with a thick sample (the length of the sample parallel to the wave

Fig. 2.4 Magnetic moment
hysteresis loops, $m(H)$ of
$Y_3Fe_5O_{12}$ compound at
295 K (room temperature).
The magnetic moment was
measured by a SQUID
magnetometer (Quantum
Design MPMS 5.5 T)
(a) This panel depicts the
magnetic moment loop in an
extended magnetic field
scale −50 kOe to 50 kOe
The saturation
magnetization has been
estimated $m_s = 27.7$
emu/gr. (b) This panel
shows the details of the
hysteresis loop in the
magnetic field
range − 1 kOe to 1 kOe.
(c) The inset shows the
measurements in the
range − 20 Oe to 20 Oe,
where a coercive field
$H_C = 14$ Oe was estimated

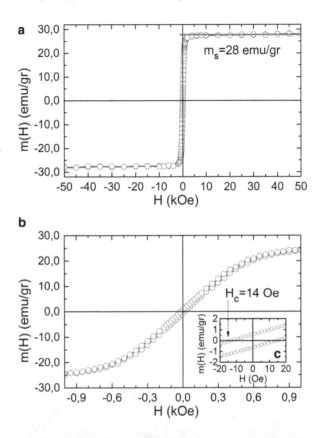

Fig. 2.5 Scanning electron
microscopy picture from the
surface of $Y_3Fe_5O_{12}$ pellet
($\times 1000$ magnification)

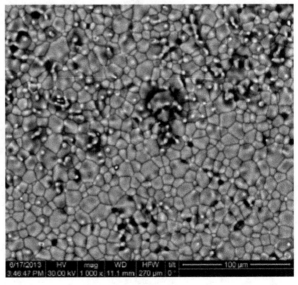

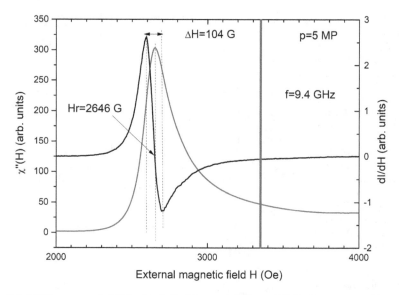

Fig. 2.6 FMR spectrum $d\chi''/dH_{dc}$ (black line) for a small piece of $Y_3Fe_5O_{12}$ compound extracted from a large bulk block. The red curve corresponds to the $\chi''(Hdc)$-curve calculated by numerical integration. The purple vertical line shows the resonance frequency for a spin with g = 2. The ferromagnetic resonance is observed in lower magnetic field in comparison to a free spin 1/2

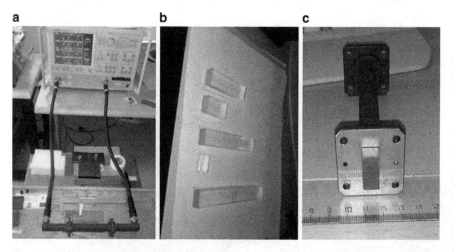

Fig. 2.7 (a) Experimental setup used to measure the scattering parameters. (b) Pieces of Plexiglas used in our measurements. (c) Wave guide section with length 128 mm. The waveguide has been loaded with $Y_3Fe_5O_{12}$

guide is larger than the wave length). In order to demonstrate the method we present measurements from three poly (methyl methacrylate) (PMMA) samples, and one from an $Y_2Fe_5O_{12}$ block. PMMA samples having lengths 7.7 mm, 32 mm, 64 mm and 128 mm, (see Fig. 2.7) were inserted inside a rectangular waveguide with cross section dimensions a $=$ 3/4$''$ $=$ 19.05 mm, b $=$ a/2 and length 128 mm. The frequency range in which the S parameters were measured, was imposed by the dimensions of the waveguide and the necessity of single mode propagation. Special attention was given to the formation of the samples to fit exactly inside the waveguide, in order to avoid the presence of undesired air gaps between the sample and the waveguide. The scattering parameters were measured using two vector network analyzers (VNA, Anritzu (see Fig. 2.7) and Rohde Schwarz) in the frequency range from 8 to 15 GHz at room temperature. A standard coaxial two-port calibration on the VNA was performed, using the TRL (Thru, Reflect, Line) calibration type. For an early account of our VNA measurements see Ref. [27].

The electric permittivity as a function of frequency was estimated by comparing the experimental measurements with the theoretically calculated, after adopting a model for the electrical permittivity. In order for this method to be applicable, the sample length must be larger than half of the wave length.

For samples with thickness lower than $\lambda/2$ the Nicholson-Ross-Weir [14] can be used to calculate the electrical permittivity.

For an isotropic non-magnetic dielectric, electrical permittivity, $\epsilon(\omega)$, could be estimated with a trial and error method, supposing a weak frequency dependence of $\epsilon(\omega)$, by comparing the theoretical calculated and experimental measured scattering parameters. The scattering parameters can be calculated in closed form for a rectangular wave guide with cross section dimensions $a\|x$-axis, and $b\|y$-axis filled with a rectangular dielectric with transverse dimensions $a \times b$ (cover all the cross section) and length d. We denote the field components with the subscript 1, 2, and 3 for the regions $z < 0$, $0 < z < d$ and $z > d$, respectively. We can satisfy the boundary conditions using only the fundamental TE_{10} mode. The electromagnetic field for $z < 0$, will be equal to the superposition of the incident and reflected waves and will be given by [17].

$$E_{y,1} = \sin\left(\frac{\pi x}{a}\right)\left(A_1^+ e^{-j\beta z} + A_1^- e^{j\beta z}\right)$$

$$H_{x,1} = -\frac{\beta}{\omega\mu_0}\sin\left(\frac{\pi x}{a}\right)\left(A_1^+ e^{-j\beta z} - A_1^- e^{j\beta z}\right) \qquad (2.8)$$

$$H_{z,1} = \frac{j\pi}{\omega\mu_0 a}\cos\left(\frac{\pi x}{a}\right)\left(A_1^+ e^{-j\beta z} + A_1^- e^{j\beta z}\right)$$

Similar equations hold inside the dielectric (region 2)

$$E_{y,2} = \sin\left(\frac{\pi x}{a}\right)\left(A_2^+ e^{-j\beta' z} + A_2^- e^{j\beta' z}\right)$$

$$H_{x,2} = -\frac{\beta'}{\omega\mu_0}\sin\left(\frac{\pi x}{a}\right)\left(A_2^+ e^{-j\beta' z} - A_2^- e^{j\beta' z}\right) \qquad (2.9)$$

$$H_{z,2} = \frac{j\pi}{\omega\mu_0 a}\cos\left(\frac{\pi x}{a}\right)\left(A_2^+ e^{-j\beta' z} + A_2^- e^{j\beta' z}\right).$$

For region 3 there exists only the wave which propagates in the right direction

$$E_{y,3} = \sin\left(\frac{\pi x}{a}\right)A_3^+ e^{-j\beta z}$$

$$H_{x,3} = -\frac{\beta}{\omega\mu_0}\sin\left(\frac{\pi x}{a}\right)A_3^+ e^{-j\beta z} \qquad (2.10)$$

$$H_{z,3} = \frac{j\pi}{\omega\mu_0 a}\cos\left(\frac{\pi}{ax}\right)A_3^+ e^{-j\beta z}$$

The propagation constants are given by the equations

$$\beta^2 = \omega^2 \epsilon_0 \mu_0 - (\pi/a)^2$$
$$\beta'^2 = \omega^2 \epsilon' \mu' - (\pi/a)^2, \qquad (2.11)$$

where $\omega = 2\pi f$ is the angular frequency. By applying the boundary conditions at the interfaces $z = 0$ and $z = d$ the scattering parameters

$$S_{11} = A_1^-/A_1^+, \qquad (2.12)$$

$$S_{21} = A_3/A_1^+, \qquad (2.13)$$

can be calculated. The results are:

$$S_{11} = \frac{\left(-1 + e^{2j\beta' d}\right)\left(\beta^2 \mu^2 - \beta'^2 \mu_0^2\right)}{\left(\beta^2 \mu^2 + \mu_0^2 \beta'^2\right)\left(-1 + e^{2j\beta' d}\right) + 2\beta\beta' \mu\mu_0\left(1 + e^{2j\beta' d}\right)} \qquad (2.14)$$

$$S_{21} = \frac{4\beta\beta' e^{j(\beta+\beta')d}\mu\mu_0}{\left(\beta^2 \mu^2 + \mu_0^2 \beta'^2\right)\left(-1 + e^{2j\beta' d}\right) + 2\beta\beta' \mu\mu_0\left(1 + e^{2j\beta' d}\right)}. \qquad (2.15)$$

The above equations associate the scattering parameters with the properties of the material inside the waveguide, namely the electric permittivity and the magnetic permeability. Obviously in order these equations to be used to compare theoretical and experimental data we must suppose reasonable frequency dependence of $\epsilon(\omega)$ and $\mu(\omega)$. A simple model is to suppose frequency independent parameters or a linear frequency variation.

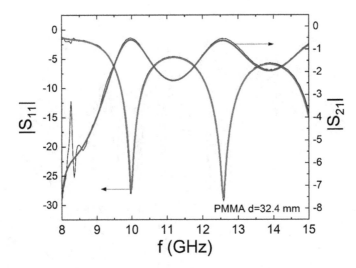

Fig. 2.8 Amplitude of the scattering parameters $|S_{21}|$ and $|S_{11}|$ versus frequency for PMMA with length 32.4 mm. Thin, black lines, represent experimental data, while the thick colored lines represent theoretically calculated data

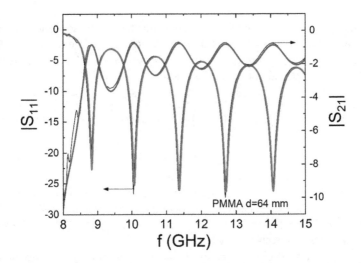

Fig. 2.9 Amplitude of the scattering parameters $|S_{21}|$ and $|S_{11}|$ versus frequency for PMMA with length 64 mm. Thin, black lines, represent experimental data, while the thick colored lines represent theoretically calculated data

Figures 2.8, 2.9, and 2.10 show the frequency variation of the experimental amplitude of the scattering parameters $|S_{11}|$ and $|S_{21}|$ for the various lengths of the specimen inside the waveguide. We observe that the $|S_{11}|$ parameter takes values near zero (≈ -25dB) at discreet frequencies, whose number increases as the length of the specimen increases. These negative peaks are attributed to a reflectionless

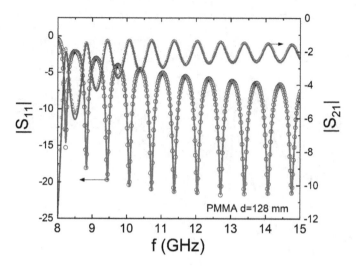

Fig. 2.10 Amplitude of the scattering parameters $|S_{21}|$ and $|S_{11}|$ versus frequency for PMMA with length 128 mm. Thin, open symbols, represent experimental data, while the thick colored lines represent theoretically calculated data

condition [15] occurring at frequencies f_p, at which the thickness of the sample is an integer multiple of the half wavelength of the electromagnetic field inside the ferrite, e.g. $d = n\lambda/2$, where n is an integer. Using this condition, $\beta = 2\pi/\lambda_g$, and the second eq. 2.11 for the propagation constant, we get

$$\varepsilon_r\mu_r = \left(\frac{c}{2f_p}\right)^2\left[\left(\frac{n}{d}\right)^2 + \left(\frac{1}{a}\right)^2\right] \tag{2.16}$$

where $c = 3 \times 10^8$ m/s is the speed of light in vacuum. Since for the PMMA $\mu_r = 1$ we can use Eq. 2.16 to get a rough estimation of its relative electrical permittivity using the frequencies where the negative peaks occur. For example, form the data shown in Figure 2.8, the reflectionless condition occurs when $f \approx 9.96$ GHz and $f = 12.6$ GHz . The Eq. 2.16 gives $\varepsilon_r(9.96$ GHz$) \approx 2.57$ and $\varepsilon_r(12.6$ GHz$) \approx 2.55$, with $n = 3$ and $n = 4$, respectively.

In the case of PMMA we observed that the permittivity is constant in the frequency range we worked. Starting from values estimated from the reflectionless condition for the electrical permittivity, with a trial and error method we succeeded an accepted coincidence between experimental and theoretical scattering parameters.

Table 2.1 displays the real and the imaginary parts of the electrical permittivity and the corresponding tanδ for the PMMA samples, with different lengths. Practically these parameters are the same for all samples. These values are also in good agreement with those reported in literature [2, 25]. In Figs. 2.8, 2.9, and 2.10 we have plotted also the theoretically calculated scattering parameters, using the real and imaginary parts of the electrical permittivity listed in Table 1.

Table 2.1 Amplitude and tanδ of the relative electric permittivity for PMMA, used to reproduce the experimental scattering parameters for pieces with different lengths, completely filling the wave guide cross section

d (mm)	ε	tanδ
32	2.56	0.008
64	2.56	0.0053
128	2.56	0.0049

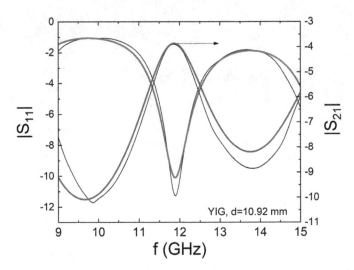

Fig. 2.11 Experimental and theoretically calculated scattering parameters S_{11} and S_{21} for bulk piece of $Y_3Fe_5O_{12}$ (a = 3/4″, b = a/2 and d = 10.95 mm)

We applied the same method for an $Y_2Fe_5O_{12}$ block measured in zero external magnetic field (see Fig. 2.11 for the measurements) supposing again that $\mu_r = 1$. The theoretical calculated scattering parameters can reproduce the experimental ones when $\varepsilon_r = 12.4$ and tanδ = 0.045.

2.5   Measurements Under a dc Magnetic Field

Here we present the results concerning the scattering parameters measured under a dc magnetic field. In this case the rectangular wave guide is loaded with sample B. This sample has dimensions a = 3/4″ in (19.05 mm), b = a/2, and d = 10.01 mm. The edge with length 10.01 mm was parallel to the z-axis of the wave guide and was located at its center. The external magnetic field was produced with two blocks of permanent magnet located at the two wide sides of the wave guide. In order a close magnetic circuit to be formed we used three thick steel sheets from magnetically soft

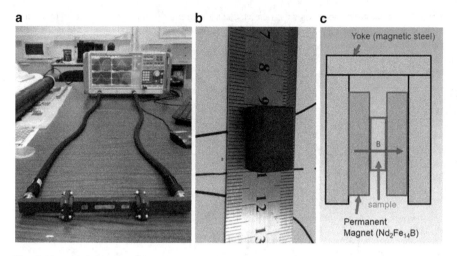

Fig. 2.12 Experimental setup for measuring the scattering parameters under a dc-magnetic filed. (a) Wave guide, (b) $Y_3Fe_5O_{12}$ sample, (c) a sketch showing the method used to produce the dc magnetic field. The magnetic field is applied along the y-axis of the wave guide

steel. This arrangement produces a dc magnetic field almost parallel to the y-axis of the wave guide (see Fig. 2.12).

Figure 2.13 shows the scattering parameters for sample B measured for $H_y = 0$, 5.55, 6.9 and 7.2 kOe. The $|S_{11}|$ parameter, under zero external magnetic field, in the studied frequency range (8–15 GHz), displays a negative peak at about 12 GHz with deep at about ≈ -10dB. Correspondingly, the transmission parameter $|S_{21}|$, forms a peak. This particular behavior is related to the reflectionless condition occurring when the thickness of the sample is an integer multiple of a half-wavelength of the field inside the material $d = \lambda_f/2 = 10.95$ mm.

In the previous section, we presented that for this $Y_3Fe_5O_{12}$ sample the zero magnetic field scattering parameters can be reproduced by asuming that $\varepsilon_r = 12.4$ and $\tan\delta = 0.045$.

For magnetic field $H_y = 5.55$ kOe, $|S_{11}|$ displays three negative peaks at about 10 GHz, 12.2 GHz and 13.8 GHz. The corresponding $|S_{21}|$ parameter forms three peaks. Similar behavior has been observed for the measurements for Hy = 6.9 and 7.2 kOe.

For a wave guide completely filled with a gyrotropic material, it is not possible for the boundary conditions at the interfaces z = 0 and z = d to be satisfied considering only the TE_{10} mode [7]. This problem can be overcome if all the TE modes are taken into account for the reflected wave in region 1, the transmitted and the reflected wave in region 2 and for transmitted wave in region 3. If, in addition, we suppose that the incident wave in region 1 is the fundamental mode$A_{1,1}^+ \sin(\pi x/a)$, then the EM-waves in the three regions can be written as follows:

Fig. 2.13 Experimental
(thin black lines) and
theoretical calculated from
the simulation (thick colored
lines) scattering parameters
S_{11} and S_{21} of $Y_3Fe_5O_{12}$
ferrite

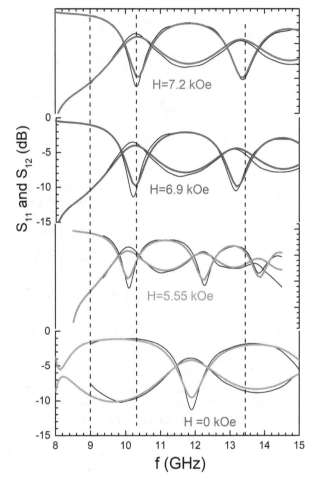

$$E_{y,1} = A_{1,1}^{+} e^{-j\beta_1 z} \sin\left(\frac{\pi x}{a}\right) + \sum_{n=1}^{\infty} A_{1,n}^{-} e^{j\beta_n z} \sin\left(\frac{n\pi x}{a}\right)$$

$$H_{x,1} = -\frac{\beta}{\omega\mu_0} A_{1,1}^{+} e^{-j\beta_1 z} \sin\left(\frac{\pi x}{a}\right) + \sum_{n=1}^{\infty} \frac{\beta_n}{\omega\mu_0} A_{1,n}^{-} e^{j\beta_n z} \sin\left(\frac{n\pi x}{a}\right)$$

$$H_{z,1} = -\frac{j}{\omega\mu_0}\left(\frac{\pi}{a}\right) A_{1,1}^{+} e^{-j\beta_1 z} \cos\left(\frac{\pi x}{a}\right) + \sum_{n=1}^{\infty} \frac{j}{\omega\mu_0}\left(\frac{n\pi}{a}\right) A_{1,n}^{-} e^{j\beta_n z} \cos\left(\frac{n\pi x}{a}\right)$$

$$(2.17)$$

for region 1 ($z < 0$))

$$E_{y,2} = \sum_{n=1}^{\infty} \sin\left(\frac{n\pi x}{a}\right)\left(A_{2,n}^+ e^{-j\beta_n' z} + A_{2,n}^- e^{j\beta_n' z}\right)$$

$$H_{x,2} = \sum_{n=1}^{\infty} \frac{\mu\beta_n'}{\omega\mu\mu_e} \sin\left(\frac{n\pi x}{a}\right)\left(-A_{2,n}^+ e^{-j\beta_n' z} + A_{2,n}^- e^{j\beta_n' z}\right)$$

$$+ \frac{-\kappa}{\omega\mu\mu_e}\frac{n\pi}{a}\cos\left(\frac{n\pi x}{a}\right)\left(A_{2,n}^+ e^{-j\beta_n' z} + A_{2,n}^- e^{j\beta_n' z}\right) \quad (2.18)$$

$$H_{z,2} = \sum_{n=1}^{\infty} \frac{jκ\beta_n'}{\omega\mu\mu_e} \sin\left(\frac{n\pi x}{a}\right)\left(A_{2,n}^+ e^{-j\beta_n' z} - A_{2,n}^- e^{j\beta_n' z}\right)$$

$$+ \frac{j\mu}{\omega\mu\mu_e}\frac{n\pi}{a}\cos\left(\frac{n\pi x}{a}\right)\left(A_{2,n}^+ e^{-j\beta_n' z} + A_{2,n}^- e^{j\beta_n' z}\right).$$

for region 2 ($0 \leq z \leq d$), and

$$E_{y,3} = \sum_{n=1}^{\infty} A_{3,n} e^{-j\beta_n z} \sin\left(\frac{n\pi x}{a}\right)$$

$$H_{x,3} = -\frac{\beta_n}{\omega\mu_0} A_{3,n} e^{-j\beta_n z} \sin\left(\frac{n\pi x}{a}\right) \quad (2.19)$$

$$H_{z,3} = -\frac{j}{\omega\mu_0}\left(\frac{n\pi}{a}\right) A_{3,n} e^{-j\beta_n z} \sin\left(\frac{n\pi x}{a}\right)$$

for region 3 ($z > d$), where $\beta_n^2 = \omega^2\epsilon_0\mu_0 - (n\pi/a)^2$ and $\beta_n'^2 = \omega^2\epsilon\mu_e - (n\pi/a)^2$. Applying the boundary conditions at $z = 0$ and $z = d$ interfaces, we get an infinite set of equations which should be satisfied. Multiplying appropriately these equations with $\cos(n\pi/a)$, $\sin(n\pi/a)$ and integrating with respect to x, from 0 to a, we finally get an infinite number of linear systems with unknowns $A_{1,n}^-$, $A_{2,n}^+$, $A_{2,n}^-$, A_3. Consequently, the scattering parameters can be expressed as an infinity series. It is obvious that this method is practically inapplicable for the estimation of the electrical permittivity and the parameters defining the magnetic permeability tensor.

It is plausible to resort to a method where the theoretical scattering parameters will be calculated by the numerical solution of the Maxwell equations or the equivalent equation obeyed by the electric field (see Eq. 2.8). To this end we use the ANSYS electromagnetic suite of programs (particularly HFSS, version 2019R3) to calculate the scattering parameters. By employing a trial and error methodology we succeeded to reproduce the experimental scattering parameters, using the same electromagnetic materials parameters for all values of the magnetic field. The scattering parameters as function of the frequency obtained by the simulation are shown in Fig. 2.13 with colored thick lines.

The simulations were made assuming a relative electrical permittivity $\epsilon_r = 12.4$ and $\tan\delta = 0.045$. These values were estimated using the methodology described in the previous section. Since the saturation magnetization was estimated from magnetization measurement $4\pi M_s = 1800$ G and the Landé factor g = 2 (the orbital moment of Fe^{3+} in $Y_3Fe_5O_{12}$ compound, is quenched) the only unknown parameter

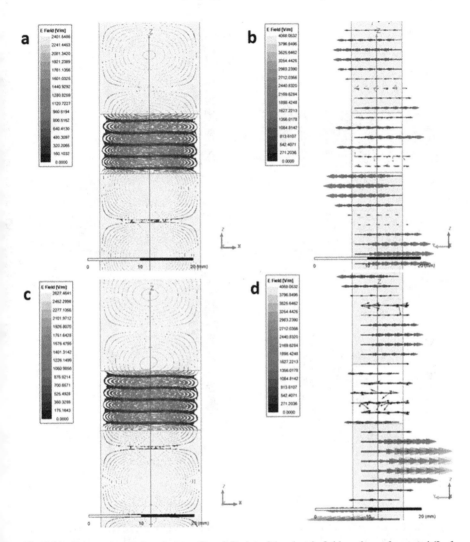

Fig. 2.14 Contour (**a** and **c**) and vector (**b** and **d**) plot of the electric field on the surface y = b/2 of the wave guide, loaded with $Y_3Fe_5O_{12}$ ferrite, under a magnetic field $H_y = 5.55$ kOe, for time phase 0° (**a** and **b**) and 45° (**c** and **d**).The frequency of the ac-electric field was f = 14.3 GHz and corresponds to the frequency where the third peak of the transmission coefficient S_{21} occurs (see Fig. 2.13 scattering parameters for $H_y = 5.55$ kOe). The ac-electric field was estimated using ANSYS HFSS (2019 R3)

needed to define the magnetic permeability tensor was the ΔH parameter (which represents phenomenologically the losses). After few trial simulations we found out that we can have an accepted coincidence between theoretical and experimental scattering parameter, when $\Delta H \approx 50$ Oe.

Figures 2.14 and 2.15 show contour and vector plots of the ac electric and magnetic fields on the surface y = b/2 of the wave guide. These plots are similar

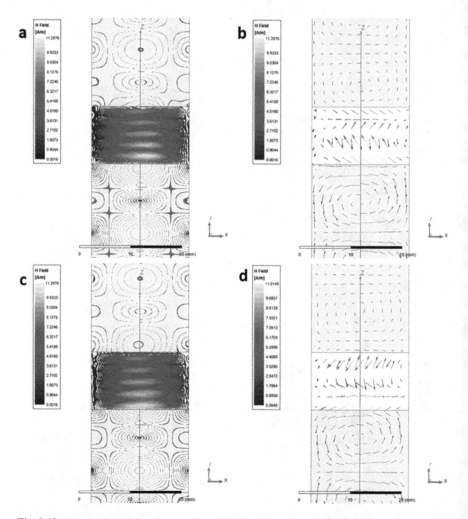

Fig. 2.15 Contour (**a** and **c**) and vector (**b** and **d**) plot of the magnetite field on the surface y = b/2 of the wave guide loaded with $Y_3Fe_5O_{12}$ ferrite, under a magnetic field H_y = 5.55 kOe for time phase 0° (**a** and **b**) and 45° (**c** and **d**). The frequency of the ac-magnetic field is f = 14.3 GHz and corresponds to the frequency where the third peak of the transmission coefficient S_{21} occurs (see Fig. 2.13 scattering parameters for H_y = 5.5 kOe). The ac-magnetic field was estimated using ANSYS HFSS (2019 R3)

(not shown here) with the ones calculated at frequencies where transmission parameter S_{21} displays a local maximum. In a first approximation the field patterns are very close to those expected from a wave guide loaded with a dielectric material, where, however, the magnetic permeability is replaced by the effective magnetic permeability parameter $\mu_e = (\mu^2 - \kappa^2)/\mu$.

The good agreement between experimental and theoretical values for the same material parameters, irrespective of the values of the external field, verifies that these parameters correspond to the average electromagnetic parameters pertinent to the whole sample. Recently, a similar methodology has been applied to characterize the electromagnetic material parameters of $Y_3Fe_5O_{12}$ ferrite [30] and ferrites in general by Thalakkatukalathil [26].

2.6 Conclusion

In the present work we have estimated the dielectric permittivity of a parallelepiped, polycrystalline $Y_3Fe_5O_{12}$ sample, and three blocks of PMMA of different length, using a broadband waveguide method. Samples with different lengths were inserted inside a rectangular waveguide WR75 and the scattering parameters were measured with a vector network analyzer (VNA). The electric permittivity as a function of frequency was estimated by comparing the experimental measurements with the theoretically calculated after adopting a linear variation of the permittivity.

Acknowledgments The present work has been partially supported by: (a) the project MIS 5002567, implemented under the "Action for the Strategic Development on the Research and Technological Sector", and (b) the project MIS 5002772, "National Infrastructure in Nanotechnology, Advanced Materials and Micro/Nanoelectronics" which is implemented under the Action "Reinforcement of the Research and Innovation Infrastructure". Both projects are funded by the Operational Programme "Competitiveness, Entrepreneurship and Innovation" (NSRF 2014–2020) and co-financed by Greece and the European Union (European Regional Development Fund).

References

1. Andreou E, Zervos T, Alexandridis AA et al (2019) Magnetodielectric materials in antenna design: exploring the potentials for reconfigurability. IEEE Antennas Propag Mag 61:29–40
2. Baker-Jarvis J, Janezic MD, Riddle BF et al (2005) Measuring the permittivity and permeability of lossy materials: solids, liquids, metals, building materials, and negative-index materials. NIST technical note 1536, Gaithersburg
3. Barzilai G, Gerosa G (1958) Modes in rectangular guides filled with magnetized ferrite. Nuovo Cimento 7:685–697
4. Barzilai G, Gerosa G (1959) Modes in rectangular guides partially filled with transversely magnetized ferrite. IRE Trans Antennas Propag 7:471
5. Carvajal-Rodriguez J (1993) Recent advances in magnetic structure determination by neutron powder diffraction. Physica B: Cond Mat 192:55
6. Collin RE (1991) Field theory of guided waves. Wiley-IEEE Press, New York
7. Epstein PS (1956) Theory of wave propagation in a gyromagnetic medium. Rev Mod Phys 28:3
8. Dionne GF (2009) Magnetic oxides. Springer, New York
9. Gurevich AG, Melkov GA (1996) Magnetization oscillations and waves. CRC Press, Boca Raton
10. Kales ML, Chait HN, Sakiotis NG (1953) A nonreciprocal microwave component. J Appl Physiol 24:816
11. Landau LD, Lifshitz EM (1935) On the theory of the dispersion of magnetic permeability in ferromagnetic bodies. Phys Zeit Sowjetunion 8:153

12. Landau LD, Lifshitz EM (1960) Electodynamics in continuous media. Pergamon Press, Oxford
13. Lax B, Button KJ (1962) Microwave ferrites and ferrimagnetics. McGraw-Hill, New York
14. Nicolson AM, Ross GF (1970) Measurement of the intrinsic properties of materials by time-domain techniques. IEEE Trans Instrument Measurement 19:377
15. Orfanidis S (2020) Electromagnetic waves and antennas. https://www.ece.rutgers.edu/~orfanidi/ewa/. Accessed 22 Mar 2020
16. Polder D (1949) On the theory of ferromagnetic resonance. London Edinb Dublin Philosoph Mag J Sci 40:300
17. Pozar DM (2011) Theory and design of ferromagnetic components. In: Pozar DM (ed) Microwave engineering, 4th edn. Wiley, New York
18. Plonsy R, Collin RE (1961) Principles and applications of electromagnetic fields. McGraw-Hill, New York
19. Queffelec P, Le Floch M, Gelin P (1999) Nonreciprocal cell for the broadband measurement of tensorial permeability of magnetized ferrites: direct problem. IEEE Trans Microwave Theory Tech 47:390
20. Queffelec P, Le Floch M, Gelin P (2000) New method for determining the permeability tensor of magnetized ferrites in a wide frequency range. IEEE Trans Microwave Theory Tech 48:1344
21. Smit J, Wijn H P J (1959) Ferrites. Philips' Technical Library, Eidhoven
22. Sodha MS, Srivastava NC (1981) Microwave propagation in ferrimagnetics. Springer, New York
23. Stancil DD, Prabhakar A (2009) Spin waves, theory and applications. Springer, New York
24. Seidel H (1957) The character of waveguide modes in gyromagnetic media. Bell Syst Tech J 36:409
25. Tanwar A, Gupta K, Singh P et al (2006) Dielectric parameters and ac conductivity of pure and doped poly (methyl methacrylate) films at microwave frequencies. Bull Mater Sci 29:397
26. Thalakkatukalathil VVK (2017) Electromagnetic modeling and characterization of anisotropic ferrite materials for microwave isolators/circulators. Dissertation, Université de Bretagne Occidentale – Brest
27. Tsiachristos I, Varouti E, Manios E et al (2014) Estimation of permeability tensor and dielectric permittivity of ferrites using a wave guide method under a dc magnetic field. EPJ Web Conference 75:06005
28. Varouti E, Devlin E, Sanakis Y et al (2020) A systematic Mössbauer spectroscopy study of $Y_3Fe_5O_{12}$ samples displaying different magnetic ac-susceptibility and electric permittivity spectra. J Magn Magn Mat 495:165881
29. Wang Y, Xu B, Zhang D et al (2019) Magneto-optical isolator based on ultra-wideband photonic crystals waveguide for 5G communication system. Crystals 9:570
30. Yao HY, Chang WC, Chang LW et al (2019) Theoretical and experimental investigation of ferrite-loaded waveguide for ferrimagnetism characterization. Prog Electromagnet Res C 90:195
31. Zervos T, Alexandridis A, Lazarakis F et al (2012) Design of a polarisation reconfigurable patch antenna using ferrimagnetic materials. IET Microwave Antennas Propag 6:158

Chapter 3
Nanopatterned Thin Films with Perpendicular Magnetic Anisotropy – Structure and Magnetism

Michał Krupiński, Yevhen Zabila, and Marta Marszalek

Abstract We describe two unconventional kinds of thin films patterning – nanosphere lithography combined with plasma etching and direct laser interference lithography. We used both techniques to create nanostructures in thin films with perpendicular magnetic anisotropy. In the first discussed case we fabricated the holes in magnetic material; in the second we generated the linear structures. We studied the magnetic properties, magnetic reversal, and the evolution of the domain pattern for matrices of Co/Pd antidots. Finally, we applied direct interference lithography to induce the transformation from disordered to ordered phase, which should result in perpendicular magnetic anisotropy of linear structures of FePt alloy and Fe/Pd multilayers.

Keywords Perpendicular magnetic anisotropy · FePt · FePd · CoPd · Nanosphere lithography patterning · Direct laser interference patterning

3.1 Introduction

Exploitation of phenomena occurring at the length scales between 1 nm and 1 μm is a key trend in modern nanochemistry, nanoelectronics, and nanobiology [1, 2]. Nanomagnetism plays a central role in the investigations of those fields; magnetic nanostructures have many applications, for example in permanent magnets, magnetic recording media, soft magnets, sensors, and materials for spin electronics [2]. The main advantage of the magnetic nanostructures is that the lowered dimensionality results in differences in the physical properties reported for bulk materials and leads to the better performance compared to naturally occurring magnetic compounds. In nanomagnetism the atomic-scale and macroscopic effects are mixed [1, 3].

M. Krupiński · Y. Zabila · M. Marszalek (✉)
Institute of Nuclear Physics Polish Academy of Sciences, Krakow, Poland
e-mail: marta.marszalek@ifj.edu.pl

© Springer Nature B.V. 2020
A. Kaidatzis et al. (eds.), *Modern Magnetic and Spintronic Materials*, NATO
Science for Peace and Security Series B: Physics and Biophysics,
https://doi.org/10.1007/978-94-024-2034-0_3

47

Magnetic nanostructures can be produced in a variety of geometries and, depending on the dimensionality, can be classified as thin films [4–6], antidots [7, 8], nanoparticles [9], nanowires [10, 11], dots [12], nanotubes [13], and nanorings [14].

For the magnetic ultra-thin films, in which the thickness is reduced to a few monolayers, the surface to volume ratio increases and the dimensionality changes from bulk 3-D to planar 2-D. The magnetic thin films have many interesting properties, such as vicinal and interface anisotropies, magnetic moment modification at surfaces and interfaces, thickness-dependent domain structure and coercivity, interlayer exchange coupling and finite-temperature ordering.

Magnetic anisotropy defined as the energy necessary to rotate magnetization direction from the easy to the hard axis is one of the most important properties of the magnetic materials and strongly affects the shape of the hysteresis loops. In thin films, the presence of interfaces and surfaces favors the in-plane easy axis of magnetization. The change of the in-plane direction to a direction perpendicular to the plane is, however, observed in many systems such as $L1_0$ ordered FePd [15, 16] and FePt [17, 18] alloys, and Co/Pd [4, 19], Co/Pt [5, 20] and Co/Ni [39, 40] multilayers. This effect is usually referred to as perpendicular magnetic anisotropy (PMA).

The Co/Pd multilayers exhibit transition from in-plane magnetic anisotropy to out-of-plane anisotropy when the Co layer thickness decreases to several atomic monolayers (typical values of Co thickness for Co/Pd multilayers are in the range of 0–25 Å [21, 22]). Despite multiple experimental and theoretical studies, the origin of the out-of-plane magnetization for the Co/Pd multilayers has not been completely understood yet. The anisotropy constant of the Co/Pd films is determined by a number of factors, such as:

- the crystal structure of the multilayer [22–24]. The largest values of perpendicular magnetic anisotropy constant K_U are observed for Co/Pd with (111) preferred orientation;
- thicknesses of the Co and Pd layers [4, 21, 25–27]. The out-of-plane anisotropy can be obtained for $d_{Co} < 20$–25 Å, the perpendicular magnetic anisotropy constant of the multilayer is more sensitive to the change of cobalt thickness;
- number of the Co/Pd bilayers. The optimum number is between 6–20 bilayers [28–31];
- buffer (seed) layer, which affects the crystalline structure of the Co/Pd multilayers [25, 32, 33];
- the quality of the interfaces [34–37] and interfacial alloying of Co and Pd.

A number of previous studies showed that the Co/Pd multilayers have perpendicular magnetic anisotropy when the Co layer is ultrathin (below 8 Å) [21, 22, 36–39], and it was also previously reported that the CoPd alloy films show perpendicular magnetic anisotropy [40, 41]. Co and Pd form spontaneously substitutionally disordered solid solutions at room temperature [34, 35, 42]. Kim and Shin [36] reported that the alloy-like character is dominant at interfaces in the typical Co/Pd multilayers and that the broken symmetry at the interfaces is not necessary for the presence of perpendicular magnetic anisotropy in the Co/Pd multilayers. In the sputtered Co/Pd multilayers, perpendicular magnetic anisotropy is caused mainly by interfacial strain anisotropy [30].

The other systems such as FePt and FePd are binary alloys with a rich phase diagram exhibiting several chemically ordered phases based on the stoichiometry of the alloy. The most interesting for PMA are the alloys with equal amounts of contributing elements.

In the bulk phase diagrams of both systems the A1 phase, corresponding to 1:1 stoichiometry ratio, is at high temperature a solid solution across the whole composition range. The crystal structure of this solid solution phase is a disordered face-centered-cubic (fcc) structure, in which Fe and Pt/Pd atoms statistically occupy the crystallographic sites. However, the fcc phase is not useful as a candidate for magnetic recording media because it is a soft magnetic phase. The high magnetic anisotropy is present only in the chemically ordered $L1_0$ alloy. Since these alloys deposited at room temperature have the fcc structure with (111) texture, it is necessary to transform the FePt/FePd alloys from the fcc structure to the $L1_0$ ordered structure. It can be done by the annealing of as-deposited alloys at high temperatures and then cooling to the room temperatures. In the Fe-Pt binary phase diagram the transformation occurs at 1300 °C, and the bulk $L1_0$-FePt alloy can be only obtained from the disordered solution by subsequent annealing. As in the Fe-Pt system, a disorder-order phase transformation in Fe-Pd also occurs but at the lower temperature of 700 °C.

The $L1_0$ structure typically has a tetragonal unit cell slightly compressed in the (001) direction, in which the Fe and Pt/Pd atoms occupy alternating (002) planes, and it is characterized by a stack of Fe and Pt/Pd atomic layers along the c direction, with the contracted lattice parameter c along the stacking direction [43]. Figure 3.1 shows the disordered fcc unit cell and the $L1_0$ structure showing the alternating stacking of (001) planes.

In the 3d metals (such as Fe) there is a strong coupling between the orbital moment and the crystal field resulting in quenching of the orbital moments. The FePt/Pd alloys with the alternating planes of Fe and Pt/Pd atoms create a highly

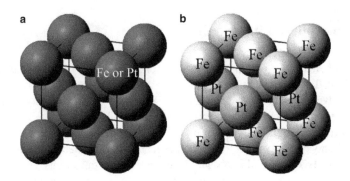

Fig. 3.1 The unit cells of (**a**) disordered fcc and (**b**) ordered fct $L1_0$ FePt alloy

anisotropic crystal, in which the spin is more tightly bound to the lattice and the magnetic properties are also highly anisotropic.

The ordered $L1_0$-FePt or FePd alloys have large magnetocrystalline anisotropy resulting from the large spin-orbit coupling of the 5d electrons. The value of magnetic anisotropy energy density K_u for the FePt and FePd alloys for thin films is similar and equals 6.0 MJm^{-3} and 1.03 MJm^{-3} for FePt and FePd, respectively [44]. The disordered alloy, on the other hand, has a much lower anisotropy of 6.0×10^{-2} MJm^{-3}.

The FePt/FePd thin films deposited directly onto an amorphous substrate tend to have the A1 fcc structure with (111) preferred orientation because the (111) plane of the disordered fcc phase is the close-packed plane and has the lowest surface energy. The transformation from the disordered A1 fcc phase to the ordered $L1_0$ superstructure is observed after annealing at temperatures larger than 500 °C. A lot of effort has been invested in the search for methods of inducing a disorder-order transformation in FePt/Pd alloys, from choosing the substrate [45–48], through doping with different impurities [49–51], to introducing the buffer layer into the system [52–55]. All these attempts required post-deposition annealing at temperatures larger than 500 °C or at least the deposition on substrates being heated to minimum 300 °C, the procedures not favorable for industrial technology.

Patterning of the magnetic thin films with PMA enhances their properties, and influences the mechanisms of demagnetization which has wide potential applications in the field of magnetic storage, sensors, radio frequency components, information processing, and magnonic crystals [56–58]. This specific interest is primarily due to the possibility to control the magnetic properties by introducing artificial nanodefects such as submicron antidots or non-magnetic inclusions ordered in large area arrays. This in turn enables tailoring of the magnetic material properties in a controlled way [59, 60].

Below we briefly describe the structure and basic properties of 3d-4d/5d systems patterned by nanosphere lithography (NSL) and direct laser interference patterning (DLIP). We applied here the power laser beam for sample annealing and, additionally benefiting from the light interference, as a method of sample patterning.

3.2 Nanosphere Lithography

The nanosphere lithography method (NSL) is one of the most successful approaches to the fabrication of nanopatterned materials. The beginnings of NSL date back to the 1980s, when Fischer and Zingsheim [61] showed that a monolayer of 10 μm polystyrene particles can be used as a mask in photolithography and can create a hexagonally ordered periodic pattern on the photoresist. The idea was quickly developed and transformed into a separate nanopatterning technique [62], which over the years has expanded into several varieties and modifications. Currently NSL includes many approaches utilizing particles made of various materials and shapes, with sizes between 10 nm and tens of micrometers. These particles can be arranged into a mask by spin coating, dip coating, self-assembly at liquid-gas and liquid-liquid interfaces, or by a combination of different techniques. The mask may by further post-processed by, for example, etching, annealing, or irradiation [63–66]. Regardless of the chosen path, the core idea remains the same: all the NSL approaches exploit the tendency of the particles to create densely packed monolayers with a long range order and then use such monolayers as a mask in the subsequent stages of nanopatterning or as a template for thin film deposition.

Over the last few decades, the NSL methods have gained wide popularity due to some key advantages. One of them is their low-cost, since the technique does not require expensive equipment as opposed to standard lithographic methods such as UV-photolithography, holographic interferometry, electron beam lithography or X-ray lithography. NSL is also able to provide structures smaller than diffraction-limited resolution defined as $\lambda/2$, which is about 100 nm for UV-photolithography. Many studies have shown that NSL can be successfully used to create sub-10 nm structures [67, 68]. In addition, it provides a higher sample throughput and a larger patterned area compared to electron beam lithography. Typically, the uniformity of the pattern can be maintained on the surface of several cm^2 and such big pattern may be obtained within several minutes. The main disadvantages of NSL are geometric restrictions. The symmetry of the obtained arrays is usually hexagonal, while the shape of the structures is round or elliptical. However, these limitations are not critical in many applications, which made the NSL a convenient fabrication method for the nanomaterials for electronics, magnonics, and optics, especially on a small laboratory scale. It has also been used for many years to tailor modern magnetic materials for magnetic storage, sensors and radio frequency components.

Here, we discuss the magnetic properties of thin films with perpendicular magnetic anisotropy patterned by NSL combined with plasma etching. We address preparation details and changes in magnetic properties for arrays of dots and antidots created in the Co/Pd multilayers, a representative system with PMA. Although our discussion concerns one material, the presented method can be directly applied to other systems exhibiting PMA such as Co/Pt, Co/Ni etc.

A general scheme of the method adopted for samples preparation is presented in Fig. 3.2. The process starts with the substrate treatment. In this particular case, Si (001) substrates covered with a native oxide layer with a thickness of about 0,5 nm were used. They were cut into pieces of 1×1 cm^2 in size, washed in organic solvents and then hydrophilized by plasma treatment in nitrogen and oxygen mixture (2:1) for 10 min. Next, the substrates were placed under water filling a large Petri dish (diameter more than 10 cm). Then polystyrene nanospheres with average diameters of 438 nm $\pm$ 10 nm were applied to the surface of water where a highly ordered hexagonal close packed monolayer of the spheres self-assembled. The monodisperse aqueous non-functionalized polystyrene (PS) suspensions were purchased from MicroParticles GmbH Berlin and in order to facilitate their spreading on the water surface they were mixed with ethanol. After a few minutes, the particles formed a compact layer, which was then crystallized by swaying of the Petri dish resulting in excitation of waves on water surface. The monolayer was than deposited on the Si substrates by slow water evaporation.

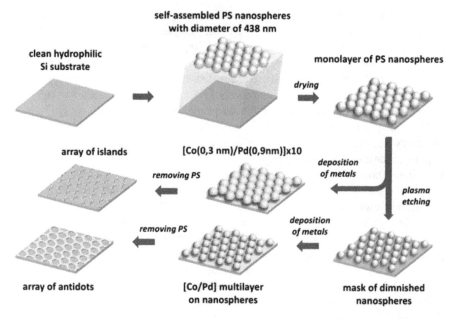

Fig. 3.2 General scheme of sample preparation using nanosphere lithography complemented by plasma etching. Period of the hexagonally ordered structures is determined by the initial diameter of the polystyrene nanoparticles and is 438 nm in this case. The plasma etching time determines the character of the obtained structures. In the case of no etching or very short etching, the size of PS particles is not reduced, and as a result an array of islands is obtained. For longer etching time, PS particles separate from each other and finally an array of holes (antidots) is formed

Next, RF-plasma etching was used, resulting in a decrease of the spheres size, but maintaining their original positions and arrangement [69]. The selection of etching time depends on the initial and final size of the nanospheres and in this particular case was ranging from 0 to 10 min. The final sphere size was inversely proportional to the plasma etching time and was chosen to be in the range of 45–100% of the initial sphere diameter. The plasma process was performed in oxygen and argon atmosphere at pressure between 0.1 and 0.2 mbar and temperature of approximately 30 °C with a chamber base pressure of 0.06 mbar. Structures obtained this way were characterized by the same period of 438 nm but with variable distances between the polystyrene spheres ranging from 0 to 240 nm.

Depending on the plasma source, the etching process may run with various etching rates and have either isotropic or anisotropic character. In order to modify the mask of nanospheres in the NSL technique, plasma etching is performed usually in an evacuated parallel-plate reactor, where the plasma is activated by a radio frequency source. In this case ions created in the reactor are directed by the local electromagnetic fields determined by parameters and shape of the reaction chamber. They are chosen so that the ions hit the PS nanoparticles perpendicular to the sample surface. This give rise to anisotropic physical etching. Consequently, spherical polystyrene spheres exposed to an oxygenated plasma undergo an anisotropic

Fig. 3.3 Scanning microscopy images taken at an angle of 30° for array of PS particles after plasma etching. The etching was carried out in argon and oxygen atmosphere under pressure of 0.15 mbar with adopted time of 7 min. As shown, the long etching results in a rough surface of nanoparticles, leading to holes (antidots) with corrugated edges

shape modification ranging from spheres to oblate spheroids. Additionally, due to polymer degradation associated with cross-linking of polymer chains, the surface of nanoparticles begins to roughen [69].

These processes are particularly visible after longed etching. Figure 3.3 presents SEM images of PS particles after 7 min of etching in argon and oxygen atmosphere under pressure of 0.15 mbar. The nanospheres are no longer spherical and resemble the shape of lens. Their surface is degraded and not flat anymore, and the edges are corrugated. This effect becomes very strong when the nanoparticle size is reduced to 60% of the initial diameter or less. As a result, the final patterned material consists of jagged and irregular structures. For this reason, long etching times should be avoided. The inability to obtain regular structures of any size for a given period is the next limitation of the presented method.

The arrays of PS particles prepared this way may be used in the subsequent stages of nanopatterning as a mask or as a template for thin film deposition. In discussed case, Pd (5 nm)/[Co (0.3 nm)/Pd (0.9 nm)]10/Pd (2 nm) multilayers were deposited on the sphere mask by sequential thermal evaporation at room temperature with evaporation rates of 0.4 nm/min for Co and 0.5 nm/min for Pd. The working pressure was in the range of 10^{-9} mbar and the film thickness was controlled *in situ* by a quartz microbalance and *ex situ* by x-ray reflectometry. After the deposition, the spheres were removed from the sample by ultrasonic assisted lift-off in toluene, leaving behind the multilayer with a hexagonal array of holes (antidot array) or triangle islands (dot array), depending on the employed plasma etching time. Mask

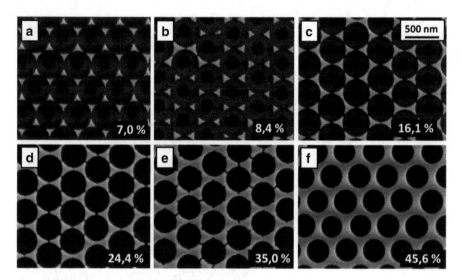

Fig. 3.4 SEM images of a Co/Pd dot and antidot array with period of 438 nm. Coverage ratios of sample surfaces with magnetic material are specified in the lower right corner in each image. Dark spots in the center of the antidots are the organic leftovers from the removed nanospheres

removal can be also done by peeling, exfoliation, or by the second plasma etching in oxygen atmosphere [59].

Lift-off completion was confirmed by inspection under a scanning electron microscope (SEM), as shown in Fig. 3.4. Such arrays can be described both by the size of the obtained structures and by the coverage of the sample surface with magnetic material. The coverage ratio for each sample may be easily determined based on brightness analysis of SEM images and depends on the size of the nanoparticles in the mask as well as on the period. In the case of non-etched spheres, dot (island) arrays were obtained (see Fig. 3.4a) with the coverage ratio of approx. 7%. As the PS particles size in the mask decreases, the coverage ratio increases up to the value close to 80% (not shown).

SEM studies show that the transition between arrays of triangular islands and antidots is not sharp. As the etching time increases, the PS particles gradually separate from each other and the gaps between them open. After reaching the percolation threshold, the system becomes continuous and antidot arrays form in which the diameter of the antidots depends on the etching time used in the earlier stage of preparation. In all cases, a large degree of long-range order is observed, which is characteristic for NSL basing on polystyrene particles assembled on water-air interface. In this particular case, highly ordered structures were obtained over a large area up to 1 cm^2.

SEM-EDX mapping images of the obtained structure are shown in Fig. 3.5. The distribution of Pd is positively correlated with that of Co and they both are associated with the regions between the holes. The signal of silicon originates mainly from the

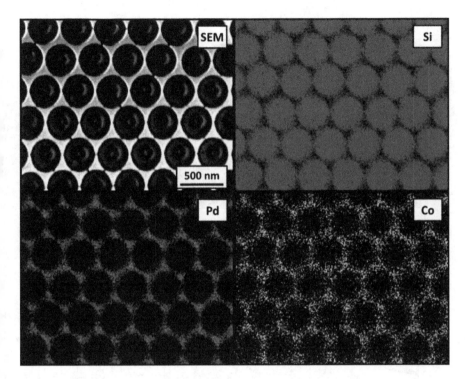

Fig. 3.5 SEM image and the corresponding EDX mapping images showing silicon, palladium and cobalt distributions in a 2D arrays of holes with period of 438 nm. Spots visible in SEM scan in the center of the antidots are the organic leftovers from the removed nanospheres

substrate exposed in the holes. The distribution of all elements reflects the topography signal and is the same for the surface of the entire sample.

Further details of the sample surface are revealed by atomic force microscopy. Figure 3.6 shows one representative AFM/MFM scan for Co/Pd array of antidots corresponding to system depicted in Fig. 3.4d. Topography scan shown in Fig. 3.6b presents a long-range hexagonal order of the obtained structures. The quality and a high long-range order of the prepared two-dimensional systems was confirmed by Fourier transform shown in the inset. Additionally, residues of nanoparticles are visible in the center of the antidots that appear as bright circles located directly on the silicon substrate. These circles are also faintly visible in some SEM images and correspond to organic residues that were not removed during the nanoparticle lift-off process. However, they are non-magnetic and are placed outside Co/Pd multilayer, and so they do not affect the properties of PMA material.

Magnetic properties of the nanopatterned Co/Pd multilayers were characterized using a SQUID magnetometer at room temperature. Hysteresis loops for each sample were measured with magnetic field in a range of ±50 kOe, large enough to saturate the magnetization, for out-of-plane and in-plane geometry. Three representative hysteresis loops for an array of separated islands (Fig. 3.7a), an array of

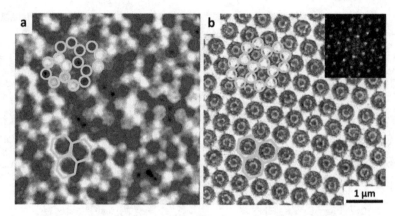

Fig. 3.6 (**a**) Magnetic domains for antidots array imaged by MFM after demagnetizing the sample (magnetic surface coverage: 24%, period: 438 nm); (**b**) The corresponding AFM topography of the same region of the sample. The inset shows the Fourier transform of the image

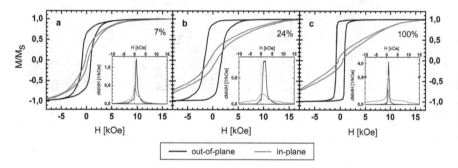

Fig. 3.7 Hysteresis loops for Co/Pd arrays with period of 438 nm. The magnetic surface coverage ratio for each sample was specified in the upper right corner of the graphs. The smallest coverage ratio corresponds to array of separated islands (**a**), the second graph presents the results for array of antidots (**b**), while the last one is for the reference flat sample (**c**). First derivatives of the lower branches for the out-of-plane and in-plane magnetization curves measured are presented in the insets

separated antidots (Fig. 3.7b), and a continuous multilayer as a reference (Fig. 3.7c) are shown in Fig. 3.7, together with first derivatives of lower magnetization branches presented in the insets. In each case the easy axis of magnetization is perpendicular to the sample surface, which means that the PMA is preserved in all patterned samples.

The highest effective magnetic anisotropy constant, determined from the area difference between the out-of-plane and in-plane hysteresis loops, was observed for flat reference sample and was 0.37 MJ/m^3. PMA for flat Co/Pd multilayer is also reflected in loop squareness, which for out-of-plane direction is high (Mr/Ms = 0.95), while for in-plane direction is significantly smaller (Mr/Ms = 0.1). The value of H$_C$

for the out-of-plane direction is 900 Oe with a saturation magnetization Ms of 800 emu/cm^3, which is in agreement with the literature data [70, 71].

PMA is preserved in all nanopatterned samples, although the arrays exhibit pronounced changes in the main magnetic parameters in comparison with non-patterned multilayer. Loop squareness for out-of-plane direction is always bigger than for in-plane geometry, however its value is smaller in comparison with the flat sample, signifying some distribution of magnetic easy axis directions. The out-of-plane squareness values range from 0.5 to 0.9 and are the lowest for the dot arrays, which is due to the increasing influence of the edge defects which locally lower the anisotropy. The decrease in anisotropy is especially visible in Fig. 3.6a, in which the shape of the loop for in-plane direction is similar to the out-of-plane loop.

During the geometrical transition from the array of dots to antidots, the width of the switching field distribution also changes. Normalized switching field distributions ($\sigma_{norm} = \sigma_{SFD}/H_C$) for out-of-plane directions are 52%, 44%, and 19% for samples with period of 438 nm and coverage ratio of 7%, 24%, and 100%, respectively. A large value obtained for the sample with the smallest coverage ratio can be explained by the diversity of size and differences in microstructure between the islands [72]. The above changes are also associated with a decrease in the effective anisotropy coefficient, which is the smallest for the array of islands and is approx. 0.1 MJ/m^3. As the size of the holes decreases, the anisotropy coefficient increases monotonically to the value characteristic for the flat multilayer.

In addition to constant anisotropy and distribution of the switching field, the coercive field is the next parameter that varies significantly after introduction of the antidots to the Co/Pd multilayers. For all samples the coercive field for in-plane geometry varies only slightly within the range of 300–500 Oe; however, for out-of-plane direction it shows pronounced changes. This is represented by the widening of the out-of-plane loops shown in Fig. 3.7. In the case of the flat non-patterned layers, its value is about 800 Oe, which remains in agreement with other studies on flat Co/Pd systems [73, 74]. For antidot arrays, its values are larger and increase as the holes size increases. A maximum coercive field of 2000 Oe is observed for the array of large antidots, which form narrow necks between neighboring holes with a width of approximately 20–30 nm. Maximum H_C value is 2.5 times bigger than for the non-patterned multilayers and may be even higher for the arrays with a smaller period [59, 60]. With further increase of the size of the antidots, the coercive field firmly decreases and reaches the lowest values for the system of separated islands.

The observed maximum of H_C and changes of K_{eff} can be understood through magnetic microscopy. Representative magnetic force microscopy (MFM) scan performed at room temperature in zero external field is shown in Fig. 3.6a. The presented arrays correspond to the maximum coercivity and were demagnetized before the measurement. Regular magnetic domains with the same shapes and sizes are present, arranged in honeycomb-like network. A single-domain state is observed where yellow and blue contrast in the MFM image corresponds to the magnetization pointing up and down, respectively. The high degree of domain order reflects the topography of the sample. Although the magnetic material is still continuous, domains constrained between antidots are magnetically separated by domain walls

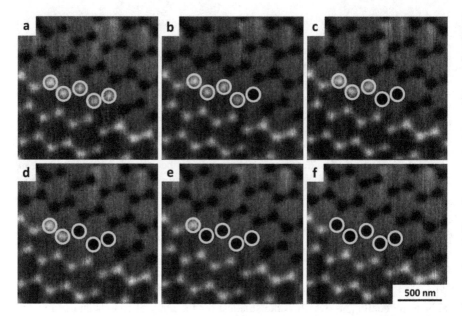

Fig. 3.8 SMRM images for antidots array (magnetic surface coverage: 24%, period: 438 nm). Scans a–f show subsequent magnetic reversal of five neighboring domains

pinned in the narrow necks between each two neighboring holes. Due to the strong pinning effect, domain walls cannot move freely through the material despite being connected through the necks.

Strong magnetic pinning of domain walls can be used to switch the domains independently. This has been shown by scanning magnetoresistive microscopy (SMRM), which is a scanning probe technique utilizing a conventional magnetic read/write head (RWH) of a hard disk drive as a sensor [74]. Its spatial magnetic resolution is in the range of 20–30 nm and unlike MFM microscopy, the scanning probe in SMRM does not generate the stray magnetic field during imaging. Its next advantage is the ability to apply local magnetic field pulses to the studied material. The size of the write pole is in the order of 100 nm × 100 nm, while the maximum available writing field of about 10 kOe. An example of magnetic imaging utilizing SMRM technique is shown in Fig. 3.8. The sample studied was the same as that shown in Fig. 3.6 and the local field option of the RWH was used to switch individual magnetic islands as demonstrated by the subsequent magnetic reversal of five adjacent regions.

Regularly spaced artificial magnetic out-of-plane domains in a continuous multilayer system are usually difficult to fabricate since the existence of domains, their shape, and regularity are usually determined by the free energy minimization and are, therefore, characteristic only for a certain thickness and the Co/Pd bilayer repetition [73]. These naturally occurring domains are usually not arranged in regular dot patterns at defined places. The patterning technique used here allowed us to overcome this restriction.

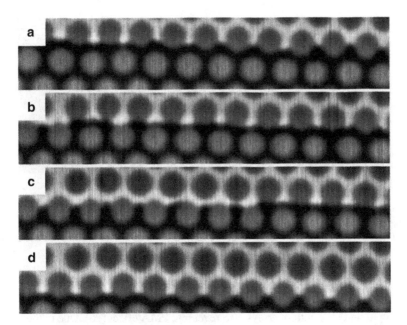

Fig. 3.9 Subsequent domain wall movement through the array (magnetic surface coverage: 55%, period: 438 nm), induced by a conventional recording head. Step size in y-direction is 50 nm

For larger amount of magnetic material between the holes, the domain wall movement becomes easier as the depinning field gets reduced [59, 75]. Figure 3.9 shows the magnetic reversal of an antidot array with antidot diameter of 310 nm, which corresponds to a coverage ratio of 55%. In the image one long horizontal domain wall can be observed between dark and bright regions corresponding to the magnetization pointing up and down, respectively. In the experiment, the domain wall was forced to move down by applying a local magnetic field from write pole of the RWH. In each step the position of the head was moved by 50 nm in down direction across the surface with sequential imaging. This has allowed us to investigate the domain wall propagation. As shown, the domain wall propagates not continuously but step-wise, being pinned and depinned at different locations in the antidot array. Such behavior is caused by the pinning occurring at the necks between two adjacent holes. When the antidot size decreases, the width of the bridges is getting bigger, which weakens the pinning effect.

The strong pinning effect of domain walls at the necks between two adjacent holes in combination with the opportunities created by SMRM can be used to create a regular lattice of magnetic domains as shown in Fig. 3.10.

For this purpose, the local magnetic field pulses with opposite polarity provided by the RWH were applied at the necks between the holes and at the area between three adjacent antidots. Period and the width of the magnetic pulses have been selected according to the geometrical dimensions of the antidot array. This resulted in an ordered domain network with domain walls localized on the necks between the

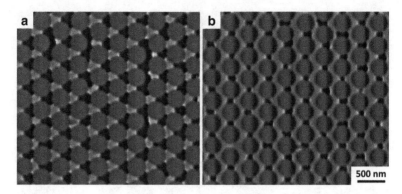

Fig. 3.10 SMRM image of an ordered magnetic domain network created by applying local magnetic field pulses (**a**) at the area between three adjacent antidots and (**b**) on the necks between two antidots. The antidot array has a magnetic surface coverage of 46% and period of 438 nm. (From M Krupinski et al. Nanotechnology 28 (2017) 085302)

holes. One neck between two holes can also support two domain walls as shown in Fig. 3.10b. Similar double walls pinned between two antidots were also observed by Grafe et al. in GdFe antidot arrays also revealing PMA effect [76]. Due to the bigger size of the magnetic domains and the zero net magnetization of the network presented in Fig. 3.10a, the system on the left-hand side is more stable and less susceptible to changes caused by an external magnetic field or temperature.

3.3 Direct Laser Interference Lithography

The direct laser interference lithography (DLIL) [77] is a method for the material surface modification, with which various interference patterns can be generated directly, permanently, and efficiently on the surfaces of different kinds of materials, such as metals, semiconductors, ceramics, or polymers [78–86]. Compared with the more common lithography methods, the novelty of laser interference structuring lies in the direct processing of the sample material itself rather than the exposure of intermediate material such as a photoresist.

Laser interference irradiation has been proven to produce periodic pattern with surface structures in the range of nano to micrometers in size. Due to the short duration of the thermal exposure by the pulse laser, the phases and morphology of deposited films can be modified with precision. Generally, this technique is used to initiate metallurgical processes such as melting, recrystallization, recovery, and the defect and phase formation in the lateral scale of the microstructure itself and with an additional long-range order given by the interference periodicity. It is therefore also called 'Laser interference metallurgy'. Because of the small thickness of layers and the quick input of the energy, the processes of melting and solidification occur in a short period of time. This enables the creation of the periodic pattern of features with a well-defined long-range order on metallic surfaces, in the scale of typical micro-structures (i.e. from the sub-micrometer level up to the micrometers). Thus, the long range ordered arrays of such pattern of phases, defects, grain sizes, textures, or stresses could be created. This could open up a considerable potential to tailor mechanical and other properties by microstructural surface functionalization.

Because planar laser waves are applied, the structuring can proceed for the macroscopic diameters of the laser beam, and consequently it should allow a fast functionalization of surfaces with dimensions of technical relevance. The physically well-defined behavior of laser light and its relatively well-known interaction with metals allows for a quantitative simulation of the experiments [87, 88].

The numerous cases of DLIL application have been described to date [88–91]. For example, the topographic structuring of the metallic multilayers, as well as the lateral formation of patterns of the B2 intermetallic compounds (NiAl and RuAl) and $L1_2$ intermetallic compound (Ni_3Al) were discussed by Mücklitz et al. [87]. In this paper, a TEM cross-section of one of the regions below the laser intensity maximum showed that the first bilayers have significant molten fractions and were deformed. The authors also showed that it is possible to locally supply enough thermal energy to restrict the formation of intermetallics in the small scale of a few microns. This scale corresponds to the interference periodicity which would allow the formation of surface composites. Another example concerns a metallic thin film sample comprised of three layers composed of Fe, Cu, and Al on a glass substrate irradiated with an interference pattern in a configuration producing a line-type energy distribution [89]. The laser fluence was high enough to melt the aluminum and copper layers at the interference maxima but the iron layer remained in the solid state. Thus, diffusive and convective exchange occurred between aluminum and copper at the energy maxima leading to the periodical alloy formation with a long-range order. The iron layer remained in the solid state at the top and acted as a protective layer, effectively preventing removal of the molten layers. The microstructural evolution of irradiated Ni/Al multi-films was further studied by Daniel et al. [90] who found that the layer interface was semicoherent. Up to a certain depth, intermetallic compounds were found in the layer interface. Surface tribological properties were optimized by combining the surface topographical patterning with the phase microstructuring [91]. A two-dimensional interference pattern was applied, and the topography and the phase microstructure of various metals were manipulated periodically.

We present here the use of DLIL patterning for materials with perpendicular magnetic anisotropy, which are good candidates for high density magnetic recording media since they have a high magnetocrystalline anisotropy energy density. However, high magnetic anisotropy exists only for the chemically ordered $L1_0$ phase. As we already discussed in introduction, directly after deposition at room temperature the FePt and FePd thin films exhibit a disordered face centered cubic (fcc) A1 structure. The different thermal procedures are required to transform the A1 structure to the $L1_0$ structure. One kind of thermal treatment leading to formation of the ordered phase is the post-deposition annealing at high temperatures. Alternatively, the alloy can be deposited onto heated substrates. The other way is the post-deposition annealing of multilayers fabricated by sequential deposition of Fe and Pt/Pd films with well-defined and carefully controlled thicknesses, providing proper stoichiometry of annealed alloys. All these methods, however, require a high temperature and a long annealing time unfavorable for manufacturing technologies. Therefore, the techniques that enable ordering at significantly reduced temperatures are urgently needed. In this chapter we describe the application of the direct laser interference lithography as an annealing and patterning technique used for the creation of arrays of FePt and FePd alloy microstructures and for the modification of their magnetic properties with the final goal of obtaining the ordered microstructures with perpendicular magnetic anisotropy. We applied the power laser beam for sample annealing and, additionally benefiting from the light interference, as a method of sample patterning.

The FePt alloys 3 and 15 nm thick were obtained by dc magnetron sputtering in the vacuum chamber with the base pressures in the order of 10^{-6} mbar or, preferably, filled with a low pressure (typically 10^{-3} to 10^{-2} mbar) of an argon gas to prevent reactions with the deposited materials. We sputtered FePt from a Fe target covered with Pt chips to obtain FePt films at 1:1 stoichiometry. The Fe/Pd multilayers preparation was done in the ultrahigh-vacuum chamber with the base pressure in the range of 10^{-8} Pa. Si(100) wafers with 100 nm thick amorphous SiO_2 layer were used as a substrate. They were ultrasonically cleaned in the acetone and ethanol and then rinsed in the deionized water before the deposition process. The [Fe(0.9 nm)/Pd (1.1 nm)] × 5 multilayers were deposited by sequential thermal evaporation at the room temperature, with the evaporation rates of 0.5 nm/min for Fe and Pd, and 0.3 nm/min for Cu. The working pressure during the sample fabrication was in the range of 10^{-6} Pa and the film thickness was controlled during the evaporation with a quartz thickness monitor.

The lithography setup was designed and mounted from the commercially available optical elements [92]. A high power pulsed Nd:YAG laser with a wavelength of 1064 nm, and pulse duration of 10 ns was used for the laser interference patterning. The energy of a single pulse ranged from 0 to 1.1 J. Since infrared radiation generated by Nd:YAG laser is invisible to the human eye, in order to allow adjustment of optical elements the additional low power laser diode emitting red light was attached at the reverse side of dielectric mirror transparent to the visible light. The beamlight of that diode reproduced the optical path of the laser beam allowing for the assembly and the adjustment of the remaining optical elements.

A schematic of the laser interference setup is shown in Fig. 3.11. The beam from the laser passes through a series of 50:50 beam splitters (BSW 11), and it is then directed to the sample surface by silver mirrors. This optical configuration results in four approximately equal-intensity beams hitting the sample. The angles of incidence can be adjusted by changing the distances between mirrors or between the sample and mirror planes. When the beams arrive at the sample surface, optical interference occurs, and local heating at the interference maxima can modify the local material properties or even partially ablate the sample, if a high intensity light is used.

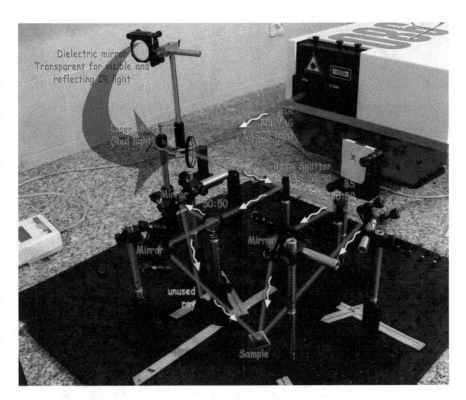

Fig. 3.11 Schematic representation of three beams interfering on the sample plane (top view). The sample plane is below the source plane at the distance z = 162 mm

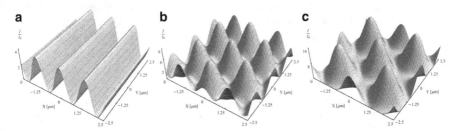

Fig. 3.12 The simulated intensities for two (**a**), three (**b**) and four (**c**) interfering beams. The intensity scale is normalized to single beam intensity I_0

For the laser interference patterning "Quantel YG980" Q-switched Nd:YAG laser with pulse duration of 10 ns, pulse energies in the near-IR wavelength of $\lambda = 1064$ nm, from 10 mJ to about 1.6 J, and repetition rates of 10 Hz was used.

Depending on the number of the laser beams and their mutual geometrical configuration the different patterning images can be obtained as we show in Fig. 3.12. The cases discussed below were obtained with a two beams system.

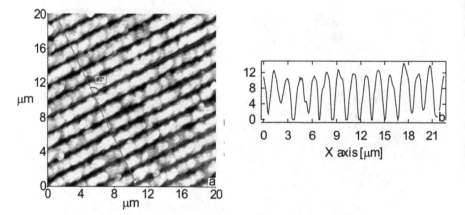

Fig. 3.13 AFM topography (**a**) and corresponding cross section (**b**) of 3 nm FePt film patterned by 175 mJ pulse

The direct laser patterning was performed on 3 nm and 15 nm disordered FePt alloys for two beam interference with the same intensities and with light wavelength of 1064 nm. The irradiations were done only with a single laser pulse 10 ns long, and the energy was adjusted during the experiments. The minimum value of laser energy (64 mJ) was the lowest value for which the changes of sample structure were observed. The maximum energy value of the laser pulse (874 mJ) was a little lower than the value at which the total removal of film was registered. These numbers were determined in preliminary experiments (not shown). The X-ray diffraction studies performed prior to irradiation indicated disordered fcc structure of FePt alloys and no magnetic contrast in the magnetic force microscope images.

Measurements of 3 nm FePt surface topography taken with AFM are presented in Fig. 3.13. The sample surface consisted of the areas covered with high stripes separated by deep valleys. The height of stripes was larger than the nominal thickness of alloy. This is demonstrated in the cross section of the surface shown in Fig. 3.13b.

We performed X-ray reflectivity measurements (not shown), which showed that for laser irradiation with small energy impulses the film density increases slightly, likely due to the short time of heating the sample. This energy, however, is not sufficient to activate recrystallization process. When bigger energy pulses are used, the film density decreases as a results of energy dissipation in the film. Increasing the laser impulse energy further again increases the film density, which then reaches the value characteristic for the as-deposited FePt alloy. This suggests that the ablation process does not occur: the density of the area neighbouring with the ablation area should not change because absorption of energy in this area is minimal. The heat absorbed by the part of sample irradiated with the maximum energy is not transferred to the neighbouring sections. We discussed earlier that the irradiation for the medium energy pulse range resulted in the decreasing of alloy density. This leads to the increase of the material volume and to the increase of the stripe height. Calculations

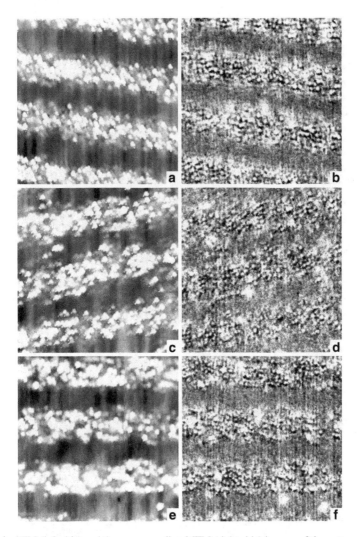

Fig. 3.14 AFM (left side) and the corresponding MFM (right side) images of the patterned FePt$_3$ $_{nm}$ film: 127 mJ/cm^2 (**a**, **b**), 275 mJ/cm^2 (**c**, **d**), 348 mJ/cm^2 (**e**, **f**). The size of each scan is 5 $\times$ 5 μm^2

based on the density obtained from X-ray reflectivity measurements and the volume of stripes calculated from integrated cross section profiles show that the entire material stays on the substrate after irradiation, which both heats up the film and separates the melted alloy in a way similar to optical tweezers.

The topography images together with magnetic contrast images for patterned FePt 3 nm alloys are shown in Fig. 3.14. The topography images show the stripes consisting of small grains. The distance between the stripes is about 1.5 μm, in agreement with values calculated from a simple equation describing the period of interference image for two beam interference. The width of stripes changed slightly

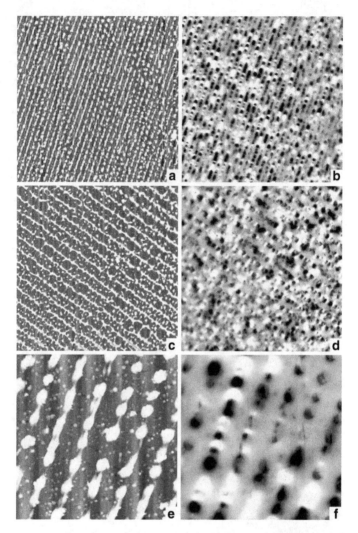

Fig. 3.15 AFM (left side) and the corresponding MFM (right side) images of the patterned FePt$_{15}$ $_{nm}$ film: 127 mJ/cm^2 50 × 50 μm^2 scan (**a, b**), 275 mJ/cm^2 25 × 25 μm^2 scan (**c, d**), 714 mJ/cm^2 10 × 10 μm^2 scan (**e, f**)

with the increasing energy of irradiation. The lack of material loss confirms that the studied energy range is below the ablation threshold.

We show the AFM/MFM images for the FePt 15 nm alloy in Fig. 3.15. We observed the continuous stripes separated by the valleys already for the smallest fluence of 127 mJ/cm^2; however, sometimes the stripes lose the continuity. The increase of the fluence to 275 mJ/cm^2 results in the further loss of continuity. Additionally, small droplets of the recrystallized material can be seen in the valleys between the stripes. For the largest fluence of 714 mJ/cm^2 the stripes are split into single islands. A better stability of the stripes observed in the AFM images of the 15 nm FePt alloy can be explained by the linear dependence between the intensity of the absorbed light and the film thickness. In the thin film, less heat is absorbed and the major portion of the radiation is absorbed by the substrate, therefore the main thermal processes do not influence the film.

The magnetization of these alloys was studied with magnetic force microscope (MFM). The MFM images of the FePt 3 nm alloy showed a weak magnetic contrast of the stripes, which correspond to the stripes observed in the topography image. The homogeneous intensity of the valleys observed in the topography of the magnetic image demonstrates the lack of the magnetic signal, whereas the stripes show the magnetic signal in which the topographic grains can be observed. With the increase of the energy the magnetic contrast in the stripes deteriorates. The observed effects are most likely related to the isotropic distribution of the magnetization direction in the magnetic phase. The magnetic images of the 15 nm FePt alloy, similar to the images of the 3 nm FePt alloy, show similarity to the objects (stripes and valleys) observed in topography. However, the stripes are divided into magnetic domains: the neighbouring regions with the maximum and minimum intensity correspond to the magnetic domains with the opposite magnetization direction. Clearly shaped domains with the in-plane magnetization are separated by domains with the out-of-plane magnetization, which is confirmed by the distinct contrast difference. Some of the neighbouring domains interact magnetostatically, which leads to the creation of a domain chain resulting in the in-plane direction of magnetization in this area. The homogeneous regions of a clearly defined contrast (an entire area white or an entire area black) most likely consist of the strong out-of-plane component of magnetization.

The changes in the topography images, which are related to the increasing energy of radiation, are reflected in the magnetic images. Figure 3.15d shows the regions with the different magnetization corresponding to the droplets in the topography valleys. Additionally, the islands observed in the topography image are reproduced in the magnetic image. For the fluence of 714 mJ/cm^2 the single grains observed in the topography have their counterpartners in the single domains with the large magnetic contrast.

The AFM/MFM images of patterned [Fe/Pd] × 5 multilayer irradiated with the different fluences are shown in Fig. 3.16a, b. The topography of the multilayers is similar to all other presented cases: it displays the valleys separated by long stripes with the round edges (Fig. 3.16c). This effect could be caused by the melting of the multilayers, despite the smaller irradiation energy than the energy used for pattering of the FePt system. The period of the obtained microstructures was 1.8 μm.

The magnetic contrast does not appear for the minimum of interference (stripe) on the [Fe/Pd]x5 multilayer irradiated with 1.08 mJ/cm^2 fluence. In fact, just the opposite happens: the valleys exhibit a distinct magnetic signal. The possible explanation is that the stripes partly kept the multilayered structure, and only a weak signal coming from the soft Fe layers was observed. The disordered FePd alloy which was fabricated in the valleys by irradiation most likely does not show the perpendicular magnetization component. The character of the FePd multilayer

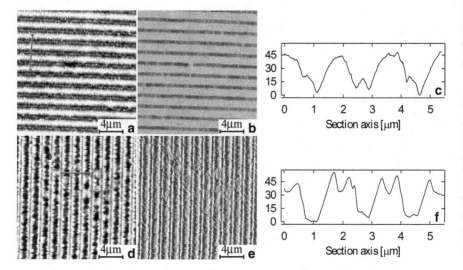

Fig. 3.16 AFM-MFM images of the [Fe/Pd] multilayers patterned with 1.08 J/cm^2 (**a, b**) and 1.30 J/cm^2 (**c, d**). Topography (**a**) and the subsequent MFM scans (**b**) are presented for 1.30 J/cm^2 pulse, while topography (**d**) and the error signal (**e**) scans are shown for 1.08 J/cm^2 fluence. The position of subsequent cross section (**c, f**) is marked in the topography image (**a, d**) as a red line. The size of each scan is 20 × 20 µm^2

irradiated with 1.3 J/cm^2 (Fig. 3.16d, e) was different than in the case described above. We did not succeed in obtaining the MFM image for this sample, therefore Fig. 3.16 d shows the AFM error signal which better displays the complex character of the microstructure created by patterning. In this case the crossection of the stripe is not squared and has a specific shape (Fig. 3.16f). The middle of the stripe appears hollow and this effect persists across the whole sample. It seems that already at this laser fluence the ablation process is initiated leading to the partial removal of the material from the valleys, which is later deposited at the edges of the stripes.

3.4 Summary

We describe here the application of two different techniques for patterning and magnetic properties modification of systems with perpendicular magnetic anisotropy. We use nanosphere lithography for patterning of Co/Pd multilayers with PMA, producing highly ordered large-scale arrays of separated dots and antidots. We show the transition between two different magnetization reversal mechanisms appearing in magnetic dot and antidot arrays. We demonstrate that the coercivity values reach a maximum for the array of antidots with a separation length close to the domain wall width. In this case, each area between three adjacent holes corresponds to a single domain configuration, which can be switched individually. On the contrary, small hole sizes and large volume of material between them result in domain wall propagation throughout the system accompanied by strong domain wall pinning at the holes.

We also apply direct laser lithography to obtain linear structures of FePt alloy and Fe/Pd multilayers. The resulting surface morphology reflects the interference pattern. We observe the periodically located stripes separated by valleys, which correspond to the minima and the maxima of laser intensity, respectively. We observe perpendicular magnetic anisotropy appears in the FePt alloy after laser irradiation. This means that laser energy led to the transformation of alloy structure from A1 to $L1_0$, but only for thicker films. This is most likely due to the poor energy dissipation in very thin film, and its fast transfer to the substrate. The irradiated Fe/Pd multilayers do not show the magnetic contrast, although they have the linear pattern. In this case the choice of laser irradiation parameters should be more careful and requires further studies.

References

1. Sellmyer D, Skomski R (2006) Advanced magnetic nanostructures. Springer, New York
2. Heinrich B, Bland JAC (2005) Ultrathin magnetic structures IV applications of nanomagnetism, 1st edn. Springer, Berlin/Heidelberg
3. Bland JAC, Heinrich B (2005) Ultrathin magnetic structures III fundamentals of nanomagnetism. Springer, Berlin/Heidelberg
4. Hu B, Amos N, Tian Y et al (2011) J Appl Phys 109:034314
5. Baltz V, Marty A, Rodmacq B et al (2007) Phys Rev B 75:014406
6. Ohtake M, Ouchi S, Kirino F (2012) J Appl Phys 111:07A708
7. Grobis M, Schulze C, Faustini M et al (2011) Appl Phys Lett 98:2011
8. Rahman MT, Shams NN, Wu YC et al (2007) Appl Phys Lett 91:132505
9. Swiatkowska-Warkocka Z, Pyatenko A, Krok F et al (2015) Sci Rep 5:9849
10. Mátéfi-Tempfli S, Mátéfi-Tempfli M, Vlad A et al (2009) J Mater Sci Mater Electron 20:249
11. Kac M, Zarzycki A, Kac S (2016) Mater Sci Eng B 211:75
12. Kanchibotla B, Pramanik S, Bandyopadhyay S (2007) In: Lyshevski SE (ed) In Nano and Molecuar electronics handbook. CRC Press, Boca Raton
13. Escrig J, Daub M, Landeros P et al (2007) Nanotechnology 18:445706
14. Adeyeye AO, Singh N (2008) J Phys D Appl Phys 41:153001
15. Iihama S, Khan M, Naganuma H et al (2015) J Magn Society Japan 39:57
16. Yabuhara O, Ohtake M, Tobari K et al (2011) Thin Solid Films 519:8359
17. Thiele J, Folks L, Toney MF et al (1998) J Appl Phys 84:5686
18. Lu Z, Walock MJ, Leclair P et al (2009) J Vac Sci Technol A 27:1067
19. Barton CW, Thomson T (2016) J Appl Phys 118:063901
20. Chowdhury P, Kulkarni PD, Krishnan M et al (2012) J Appl Phys 112:023912
21. den Broeder FJA, Hoving W, Bloemen PJH (1991) J Magn Magn Mater 93:562
22. Engel BN, England CD, van Leeuwen RA (1991) Phys Rev Lett 67:1910
23. Shaw JM, Nembach HT, Silva TJ et al (2009) Phys Rev B 80:184419
24. Barnes JR, O'Shea SJ, Welland ME et al (1994) J Appl Phys 76:2974
25. Shaw JM, Rippard WH, Russek SE et al (2007) J Appl Phys 101:023909
26. Nemoto H, Hosoe Y (2005) J Appl Phys 97:10J109
27. Gottwald M, Lee K, Kan JJ et al (2013) Appl Phys Lett 102:052405
28. Tadisina ZR, Natarajarathinam A, Clark BD et al (2010) J Appl Phys 107:09C703
29. Sbiaa R, Ranjbar M, Åkerman J (2015) J Appl Phys 117:17C102

30. Rozatian ASH, Marrows CH, Hase TPA et al (2005) J Phys Condens Matter 17:3759
31. Kawaji J, Asahi T, Onoue T et al (2002) J Magn Magn Mater 251:220
32. Kim S, Lee S, Kim J et al (2011) J Appl Phys 109:109
33. Onoue T, Asahi T, Kuramochi K et al (2001) J Magn Magn Mater 235:40
34. Kim SK, Chernov VA, Koo YM (1997) J Magn Magn Mater 170:L7
35. Kim S, Koo Y, Chernov V (2000) Phys Rev B 62:3025
36. Kim SK, Shin SC (2001) J Appl Phys 89:3055
37. Carrey J, Berkowitz AE, Egelhoff WE (2003) Appl Phys Lett 83:5259
38. Johnson MT, Bloemen PJH, den Broeder FJA (1996) Reports Prog Phys 59:1409
39. Carcia PF (1998) J Appl Phys 63:5066
40. Hiroshi T (1993) Jpn J Appl Phys 32:L1328
41. Hashimoto S, Ochiai Y, Aso K (1989) J Appl Phys 66:4909
42. Kohlhepp JT, Strijkers GJ, Wieldraaijer H et al (2002) Phys Stat Solidi A 189:701
43. Pearson WB (1958) A handbook of lattice Spacings and structures of metals and alloys. Pergamon Press, London
44. Aitchison PR, Chapman JN, Gehanno V et al (2001) J Magn Magn Mat 223:138
45. Carbucicchio M, Ciprian R, Palombarini G (2010) J Magn Magn Mat 322:1307
46. Weil DH, Yao YD (2009) Appl Phys Lett 95:172503
47. Lukaszew RA, Cebollada A, Clavero C (2006) Physica B 384:15
48. Jeong S, Hsu YN, Laughlin DE et al (2001) IEEE Trans Magne 37:1299
49. Polit A, Makarov D, Brombacher C et al (2015) J Magn Magn Mat 381:316
50. Kitakami O, Shimada Y, Oikawa K et al (2001) Appl Phys Lett 78:1104
51. Albrecht M, Brombacher C (2013) Phys Stat Solidi A 210:1272
52. Bourgognon C, Tatarenko S, Cibert J et al (2000) App Phys Lett 76:1455
53. Yang E, Laughlin DE, Zhu JG (2010) IEEE Trans Magn 46:2446
54. Yang E, Laughlin DE (2008) J Appl Phys 104:023904
55. Xu YF, Chen JS, Wang JP (2002) Appl Phys Lett 80:3325
56. Unal AA, Valencia S, Radu F et al (2016) Phys Rev Appl 5:064007
57. Kruglyak VV, Demokritov SO, Grundler D (2010) J Phys D Appl Phys 43:264001
58. Krawczyk M, Grundler D (2014) J Phys: Cond Mat 26:123202
59. Krupinski M, Mitin D, Zarzycki A et al (2017) Nanotechnology 28:085302
60. Krupinski M, Sobieszczyk P, Zieliński P et al (2019) Sci Rep 9:13276
61. Fischer UC, Zingsheim HP (1981) J Vac Sci Technol 19:881
62. Deckman HW, Dunsmuir JH (1982) Appl Phys Lett 41:377
63. Haynes CL, van Duyne RP (2001) J Phys Chem B 105:5599
64. Zhang X, Whitney AV, Zhao J et al (2006) J Nanosci Nanotech 6:1920
65. Zhao X, Wen J, Li L et al (2019) J Appl Phys 126:141101
66. Ai B, Yu Y, Möhwald H et al (2014) Adv Coll Interf Sci 206:5
67. Wang H, Levin CS, Halas NJ (2005) J Am Chem Soc 127:14992
68. Ji D, Li T, Fuchs H (2017) Adv Electron Mater 3:1600348
69. Akinoglu EM, Morfa AJ, Giersig M (2014) Langmuir 30:12354
70. Hauet T, Hellwig O, Park SH et al (2011) Appl Phys Lett 98:172506
71. Hu G, Thomson T, Rettner CT et al (2005) J Appl Phys 97:10J702
72. Lau JW, McMichael RD, Chung SH et al (2008) Appl Phys Lett 92:012506
73. Sbiaa R, Bilin Z, Ranjbar M et al (2010) J Appl Phys 107:103901
74. Mitin D, Grobis M, Albrecht M (2016) Rev Sci Instrum 87:023703
75. Lee J, Brombacher C, Fidler J et al (2011) Appl Phys Lett 99:062505
76. Grafe J, Weigand M, Trager N et al (2016) Phys Rev B 93:104421
77. Kim DY, Tripathy SK, Ki L et al (1995) Appl Phys Lett 66:1166
78. Heintze M, Santos PV, Nebel CE et al (1994) Appl Phys Lett 64:3148
79. Phillips HM, Callahan DL, Sauerbrey R et al (1991) Appl Phys Lett 58:2761
80. Ilcisin KJ, Fedosejev R (1987) Appl Opt 26:396
81. Kelly MK, Rogg J, Nebel CE (1998) Phys Stat Sol A 166:651

82. Nebel CE, Christiansen S, Strunk HP (1998) Phys Stat Sol A 166:667
83. Fukumura H, Ujii H, Banjo H et al (1998) Appl Surf Sci 127:761
84. Fukumura H, Kohji Y, Nagasawa K et al (1994) J Am Chem Soc 116:10304
85. Shoji S, Kawata S (2000) Appl Phys Lett 76:2668
86. Lorens M, Zabila Y, Krupinski M et al (2012) Acta Physica Pol A 121:543
87. Lasagni A, Mücklich F (2005) Appl Surf Sci 240:214
88. Mücklich F, Lasagni A, Daniel C (2005) Intermetallics 13:437
89. Lasagni A, Holzapfel C, Weirich T et al (2007) Appl Surf Sci 253:8070
90. Daniel C, Mücklich F (2005) Appl Surf Sci 242:140
91. Daniel C, Mücklich F, Liu Z (2003) Appl Surf Sci 208:317
92. Zabila Y, Perzanowski M, Dobrowolska A et al (2009) Acta Physica Pol A 115:591

Chapter 4
L1$_0$ Ordered Thin Films for Spintronic and Permanent Magnet Applications

Arsen Hafarov, Oleksandr Prokopenko, Serhii Sidorenko, Denys Makarov, and Igor Vladymyrskyi

Abstract Materials with strong perpendicular magnetic anisotropy (PMA) are fundamentally appealing and also relevant for numerous applications especially considering their practical relevance for the enhancement of the energy product for thin film based permanent magnets and realization of energy efficient and miniaturized spintronic devices. In contrast to materials exhibiting PMA due to surface anisotropy, these applications would benefit from thin films where PMA stems from a strong uniaxial magnetocrystalline anisotropy (K_u). In this regard, magnetic thin films with chemically ordered $L1_0$ structure, representing alternation of A and B atomic planes along the c direction, are considered as most promising due to the high K_u values and finely tunable magnetic properties. Typical representatives of $L1_0$ structures are ordered binary phases, e.g. FePt, FePd, MnAl, MnGa, or NiFe. In the case when the c axes of the $L1_0$ structure is normal to the film plane, remarkably strong PMA can be achieved. Another important property of $L1_0$ structures is their thermodynamic stability providing resistance of corresponding devices against thermal processing. Here, we will review the application prospects of $L1_0$ ordered magnetic thin films for spintronic and permanent magnet technologies.

Keywords Thin films · Permanent magnet · Spintronic · MTJ · MRAM

A. Hafarov (✉) · S. Sidorenko · I. Vladymyrskyi
Metal Physics Department, National Technical University of Ukraine "Igor Sikorsky Kyiv Polytechnic Institute", Kyiv, Ukraine
e-mail: hafarov@kpm.kpi.ua

O. Prokopenko
Faculty of Radio Physics, Electronics and Computer Systems, Taras Shevchenko National University of Kyiv, Kyiv, Ukraine

D. Makarov
Helmholtz-Zentrum Dresden-Rossendorf e.V, Institute of Ion Beam Physics and Materials Research, Dresden, Germany

© Springer Nature B.V. 2020
A. Kaidatzis et al. (eds.), *Modern Magnetic and Spintronic Materials*, NATO Science for Peace and Security Series B: Physics and Biophysics,
https://doi.org/10.1007/978-94-024-2034-0_4

4.1   Introduction

Historically, magnetic materials are broadly applied in numerous technologies ranging from navigation, electrical machines and drives through data storage all the way to medicine and consumer electronics. There are two major research fields contributing to the realization of devices for these applications. Those are permanent magnets and spintronic technologies (Fig. 4.1). Permanent magnets are vital components of acoustic transducers, motors, generators, magneto-mechanical devices, and magnetic field and imaging systems [58, 62, 92]. Thin films of $L1_0$ ordered magnetic materials are considered as promising to enhance the energy product of rare-earth-free permanent magnets, which represent a measurement for the maximum amount of magnetic energy stored in a magnet. Spintronics offer solutions for novel devices [7, 11, 53, 69, 89], e.g. random access memories, microwave detectors and spintronic diodes, some of which already made their way to the market [2, 89]. The challenges yet to be addressed in spintronics, are lowering the power consumption, enhancing the recording density for memory devices, as well as increasing the output power of radiation emitters and detectors. CoFeB ferromagnetic thin films, which are a typical component of state-of-the-art spintronic layer stacks including magnetic tunnel junctions, face difficulties related to the physical origin of the PMA in the layers (interfacial origin), relatively large values of the saturation magnetization and Gilbert damping constant [2, 69, 89]. To overcome the emergent limitations on the material side, one of the promising approaches is to use

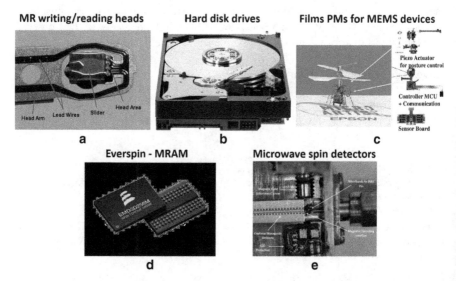

Fig. 4.1 Representative application examples of hard magnetic materials: (**a**) MR writing/reading heads. (**b**) hard drive disks. (**c**) thin films for MEMS devices. (**d**) MRAM devices. (**e**) microwave spin detectors

Fig. 4.2 TEM images of the FePt thin film grown on MgO (001) at 350 °C. (**a**) Lattice plane image showing a long-range periodic structure composed of alternating Fe (bright) and Pt (dark) planes parallel to the film plane. (**b**) Selected area diffraction pattern of $L1_0$ ordered FePt grown epitaxially on MgO. (**c**) Antiphase boundaries in the FePt film. (**d**) Atomic resolution TEM micrograph showing the evidence of dislocations at the interface between FePt and MgO. (Reproduced from [45], with the permission of AIP Publishing)

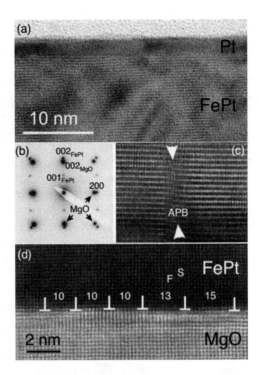

thin films possessing strong PMA due to uniaxial magnetocrystalline anisotropy and tunable other magnetic properties including saturation magnetization and damping. These requirements can be fulfilled relying on $L1_0$ chemically ordered alloy thin films.

Magnetic materials with chemically ordered $L1_0$ structure represent face centered tetragonal phase with ordered alternation of A and B atomic planes along the c axis of the lattice as shown in Fig. 4.2. Such phase could be formed in numerous binary systems including those revealing strong PMA, e.g. FePt, FePd, CoPt, MnAl, MnGa, FeNi. For instance, this property allows to implement $L1_0$-FePt ordered thin films as a medium of heat assisted magnetic recording with extremely high recording density of 1.4 Tb/in^2 [42, 106].

Here, we will focus on several representatives of the $L1_0$ materials family, which hold great potential for spintronics and permanent magnets applications. Those materials are $L1_0$-MnAl, $L1_0$-MnGa and $L1_0$-FePt. The review is organized as follows: first, structural and magnetic properties of Mn-based $L1_0$ ordered materials will be discussed. Regarding magnetic and structural properties of $L1_0$-FePt we refer to numerous reviews [5, 49, 106]. Then, applications of these alloys in permanent magnet technologies (including exchange spring coupled magnets), as well as devices based on magnetic tunnel junctions (random access memory, spin transfer torque nano-oscillators and nano-diodes) will be addressed.

4.2 Structural and Magnetic Properties of Mn-Based $L1_0$ Ordered Thin Films

$L1_0$-MnAl Thin Films $L1_0$ phase of the Mn–Al binary alloy is a well-known ferromagnet, which exhibits excellent hard magnetic properties, yet contains neither ferromagnetic elements such as Fe, Co, and Ni nor noble metals such as Pt. The theoretical predictions regarding magnetic properties of Mn-Al alloys report high saturation magnetization of 600 emu/cm^3, Curie temperature of 650 K, and maximum energy product $(BH)_{max}$ of 12.4 MGOe [32]. Remarkably high values of magnetocrystalline anisotropy constant of up to 15×10^6 erg/cm^3 were experimentally demonstrated [68, 71, 86]. Furthermore, the saturation magnetization of $L1_0$-MnAl alloys can be tuned in a wide range (40–500 emu/cm^3) [17, 71, 86].

The possibility to fabricate $L1_0$-MnAl films with controllable magnetic properties was shown in 2000s. For instance, Z.C. Yan et al. reported that structural and magnetic properties of sputtered MnAl films can be easily modified by varying the layers thicknesses, deposition temperature and post-annealing conditions [111]. By applying magnetic field during the annealing process, it is possible to tune magnetization of MnAl thin films. For example, an external magnetic field of 15 kOe allows to reduce drastically the saturation magnetization (M_s) value from 248 to 53 emu/cm^3 (about 5 times) while keeping the coercive field nearly constant [17].

These approaches to tailor magnetic properties of MnAl thin films are typically applied in recent studies. The use of MgO seed layer for the growth of 10–50 nm thick MnAl alloy films followed by their annealing in an applied magnetic field of 4 kOe allows to achieve samples with coercivity $H_c \sim 8$ kOe and anisotropy constant $K_u \sim 6.5 \times 10^6$ erg/cm^3 [25, 27]. Even more attractive hard magnetic properties of MnAl thin films were achieved in the case when TiN underlayer was used. MnAl (10–130 nm)/TiN(10 nm)/Si stacks show both high H_c of 12 kOe and K_u of 1.0×10^7 erg/cm^3 as well as a relatively small M_s of 250 emu/cm^3 [25, 27].

Molecular beam epitaxy on GaAs(001) substrates were demonstrated as another promising approach to form MnAl films with a pronounced PMA [68]. Such films exhibit high coercivity of 10.7 kOe and the energy product of 4.44 MGOe. Moreover, it was shown that the M_s value can be easily tuned in a wide range (48–361 emu/cm^3) by controlling the heat treatment parameters [67, 68]. The possibility to achieve a pronounced PMA accompanied with a small saturation magnetization in MnAl films makes them attractive for spintronic devices.

$L1_0$-MnGa Thin Films The $L1_0$ ordered MnGa compound possesses bulk PMA constant as high as 21.7×10^6 erg/cm^3 [55, 119]. The magnetization of MnGa can be varied from 27.3 to 270.5 emu/cm^3 by adjusting the film composition and growth conditions [55]. Calculations show its ultralow damping constant of about 0.0003, which could be useful in spintronic applications for the realization of low power consumption devices. $L1_0$-MnGa films with high coercivity (~10 kOe) can be formed on SiO$_2$/Si substrates [16]. Textured $L1_0$-MnGa thin films can be prepared on MgO substrates using electron beam evaporation. 100 nm thick films reveal a

large magnetocrystalline anisotropy ($\sim$10 $\times$ 10^6 erg/cm^3) and the saturation magnetization of 470 emu/cm^3 [99]. A decrease of the film thickness to 5 nm leads to lowering of the saturation magnetization to about 300 emu/cm^3, still maintaining a high value of magnetocrystalline anisotropy [99]. Thin MnGa films with a thickness from 1 to 30 nm exhibiting strong PMA could be also formed on Si substrates using MgO(10 nm), Cr(40 nm) and CoGa(30 nm) buffer layers. Even ultrathin films with a thickness of 2–3 nm were found to reveal the PMA of up to 5 $\times$ 10^6 erg/cm^3 and the saturation magnetization of $\sim$200 emu/cm^3 [70]. Fe was also used as a buffer layer for the formation of 20 nm thick MnGa films resulting in a low surface roughness of 0.93 nm and strong PMA of 9.3 $\times$ 10^6 erg/cm^3 [88].

4.3 L1$_0$ Ordered Thin Films Application in Permanent Magnet Technologies

The primary figure of merit of PMs is the maximum energy product $(BH)_{max}$, which provides the density of the magnetic flux that can be stored in a magnet. The energy product of any magnetic material directly depends on its saturation magnetization and coercivity. During the last decades $(BH)_{max}$ significantly increased from 1 MGOe for steel magnets to about 50 MGOe for sintered Nd$_2$Fe$_{14}$B magnets. The evolution of $(BH)_{max}$ values during the last century is given in Fig. 4.3 [62].

Most widely used PMs contain rare earth (RE) elements as Nd and Dy. Beside issues related to the geographical origin of REs supply, the use of such elements also causes environmental concerns, related to the manufacturing process of RE magnets. Thus, there is an intensive search for new RE-free materials with attractive magnetic

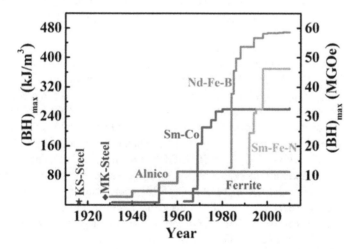

Fig. 4.3 Evolution of the performance of permanent magnets. (Reprinted from [62], Copyright (2020), with permission from Elsevier)

properties for PM applications. New magnets are expected to bridge the gap between NdFeB-based PMs (>50 MGOe of $(BH)_{max}$) and currently used RE-free PMs based on Ferrites and Alnicos (<10 MGOe of $(BH)_{max}$).

It is established that the magnetic behavior of PMs strongly depends on the structural and microstructural properties of the materials. Structure could be successfully affected by tuning fabrication parameters [12, 13, 30], post-processing treatment conditions [72, 94], alloying [14, 19, 38, 59, 109], application of external fields [17, 25]. One of the most promising approaches relies on nanostructuring magnets [86]. A clear benefit of this approach lies in the fact that the transition from a multi- to single-domain state in a magnetic nanoobject is accompanied with a sharp increase in the coercivity [22], which has a direct influence on the energy product. Another promising route is related to utilizing the exchange spring coupling (ESC) effect, which could be effectively used in the case of nanoscale magnetic systems [18, 37].

The essence of the ESC approach is in the combination of an exchange coupled soft and hard magnetic thin films in a single architecture. If the magnetic parameters in the composite are properly adjusted [40, 41, 52, 92, 95, 96], the magnetization of such a structure can be tailored for the entire layer stack but not for the soft or hard layers separately [37]. In this case, the resulting M_s of the composite will be given by a magnetically soft material (which is typically chosen to have a large Ms) while the coercive field is determined by the one of the hard magnetic material (which is typically chosen to have a large H_c). Since the $(BH)_{max}$ is determined by the M_s and H_c, ESC can lead to a substantial enhancement of the maximum energy product. It is worth noting that ESC systems are intensively explored for spintronics and information storage.

Although the ESC approach can be applied even for bulk PMs, it is exceptionally well suited for thin film based PMs as a smaller fraction of the hard magnetic phase and larger amount of the soft magnetic one can be used. Thus, larger magnetization and respectively higher $(BH)_{max}$ values can be obtained [92]. Therefore, different nanostructures including nanostructured thin films, e.g. nano-dots, nano-wires possess strong potential for high performance PMs.

In this respect, $L1_0$ chemically ordered hard magnetic thin films, in particular Fe-based $L1_0$-ordered thin films like FePt and FeNi, are of great interest. For example, using full potential linearized augmented plane wave method, it was shown theoretically that a very large energy product of 65 MGOe can be reached in FeCo/FePt exchange coupled thin films. Later G. Giannopoulos et al. [21] have experimentally shown the possibility to create magnetron sputtered $Fe_{55}Co_{45}$/FePt ESC-based PMs with $(BH)_{max}$ of 50 MGOe. Layered thin films with different number of FeCo and FePt monolayers (ML) were investigated as shown in Fig. 4.4.

It was concluded that FePt(7 ML)/FeCo(5 ML) stack shows strongest exchange coupling and the $(BH)_{max}$ value that caused by a substantial increase of the M_s (Fig. 4.5) [21]. Recently, FeCo/FePt composites with a large energy product were fabricated on Si substrates by G. Vashist et al. [103]. Two sets of samples were deposited via magnetron sputtering: FeCo(12 nm)/$Fe_{55}Pt_{45}$(27 nm)/Si and FeCo (6 nm)/$Fe_{55}Pt_{45}$(27 nm)/Si. In this case, the $(BH)_{max}$ of 47 MGOe was achieved

Fig. 4.4 A schematic illustration of (**a**) 2 ML Fe-Co, (**b**) 5 ML Fe-Co, and (**c**) 10 ML Fe-Co on 4 ML FePt substrate. (Reproduced from [21], with the permission of AIP Publishing)

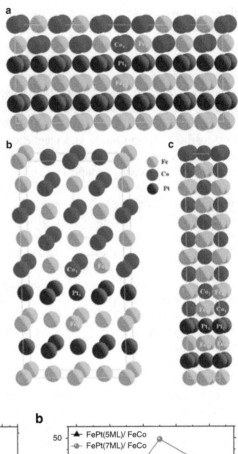

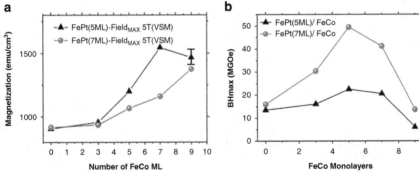

Fig. 4.5 Magnetization and maximum energy product of FePt/FeCo thin films depending on the thickness of the FeCo layer. (Reproduced from [21], with the permission of AIP Publishing)

for films with a 6 nm thick FeCo layer. An increase of the FeCo layer thickness leads to a strong degradation of the exchange coupling strength [103].

The main reason why FePt thin films in combination with FeCo show such high values of the energy product lies in a large magnetization of FeCo (~2000 emu/cm^3) [46] and strong magnetocrystalline anisotropy as well as coercivity of $L1_0$-FePt.

However, even higher $(BH)_{max}$ value was achieved using $A1$-FePt as a soft phase in combination with $L1_0$-FePt. In this respect, Y. Liu et al. produced an exchange spring $L1_0$-FePt/$A1$-FePt nanocomposite with the energy product value of 54 MGOe by varying the Fe content in FePt thin films [47]. The same group later reported experimental and theoretical results obtained for FePt/Fe exchange coupled system, where the $(BH)_{max}$ of more than 60 MGOe was achieved [48].

O. Crisan et al. showed that it is possible to form exchange spring nanocomposites based on annealed FeMnPt ternary alloy films [6]. The analysis of the Mossbauer spectroscopy and magnetic properties confirmed the existence of both hard magnetic $L1_0$-FePt and soft magnetic $L1_0$-FeMnPt phases, allowing for the formation of an exchange coupled system.

These works show that thin film PMs based on $L1_0$-FePt phase can provide $(BH)_{max}$ values comparable with bulk RE PMs. However, the cost of Pt imposes some limitations on manufacturing of such PMs. Therefore, $L1_0$-MnAl-based thin films are intensively studied with respect to their applicability as RE-free PMs. In this respect, bulk MnAl-based PMs were proven to reveal relatively large $(BH)_{max}$ values of up to 10 MGOe [15, 19, 38, 59, 109]. Still, independent of various approaches applied to improve the PM performance of bulk MnAl, including doping of material with C or B [14, 19, 38, 59, 109], using different sample fabrication and processing techniques [72, 94], the theoretically predicted values (up to 14 MGOe) of the energy product for bulk MnAl alloys are not achieved experimentally [3]. On the other hand, rather promising results were recently demonstrated using nanosized MnAl-based PMs.

Exchange spring coupling was achieved in MnAl(60 nm)/FeCo(4; 8; 12 nm) thin films by Dang et al. [8]. The samples were prepared by dc magnetron sputtering. The MnAl layer was prepared by annealing of [Mn(x nm)/Al(y nm)]$_n$ multilayers at 673 K for 1 h. Then, FeCo layer was sputtered on MnAl at room temperature. An appropriate exchange coupling between MnAl and FeCo was observed when the thickness of the FeCo layer was not larger than 8 nm as clearly can be seen in Henkel plots in Fig. 4.6d. The stack with 8 nm thick FeCo layer reveals the highest $(BH)_{max}$ value of 5 MGOe (Fig. 4.6c). Other soft magnetic layers including Co thin films exchange coupled to MnAl were studied as well. In this respect, J. You et al. [114] investigated MnAl/Co stacks. A MnAl layer with the thickness of 60 nm was deposited at 773 K by dc magnetron sputtering onto an MgO seed layer prepared on an MgO substrate. The thickness of the Co layer was varied (2; 4; 6 nm). The growth of Co was done using plasma enhanced atomic layer deposition. The efficient magnetic exchange coupling was observed when the thickness of the Co layer was 2 and 4 nm. The samples revealed the energy product of up to 3.5 MGOe.

Another approach to influence the PM performance of MnAl-based composites was proposed by T. Sato et al. [86]. $L1_0$-MnAl films with partial Mn substitution by 3d transition metals (Fe, Co, Ni, and Cu) were investigated. The use of Ni and Fe resulted in an increase in the saturation magnetization from 360 to 400 emu/cm^3. However, in the case of Ni, the increase of M_s is accompanied with the decrease of the coercive field. At the same time, the same value of H_c of about 5 kOe was preserved when Fe was used.

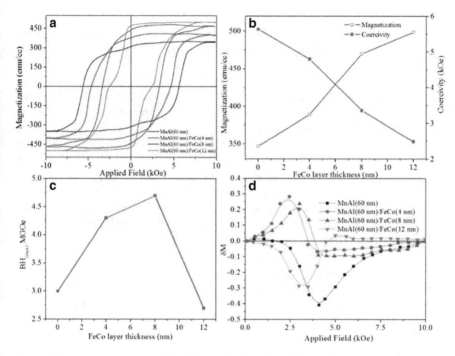

Fig. 4.6 Magnetic properties of MnAl/FeCo thin films depending on the thickness of the FeCo layer (**a**). In-plane magnetic hysteresis loops of MnAl/FeCo thin films; (**b**). Magnetization and coercivity of MnAl/FeCo films depending on FeCo layer thickness; (**c**). Energy product of MnBi and MnAl/FeCo thin films; (**d**). Henkel plots of MnBi and MnAl/FeCo films. (Reprinted from [8], Copyright (2020), with permission from Elsevier)

With respect to the PM performance of $L1_0$-MnGa thin films, Sabet et al. showed experimentally and theoretically that the exchange coupling between the hard $L1_0$-MnGa and soft 2-nm-thick FeCo magnetic layers allows to increase the saturation magnetization for 20% compared to pure MnGa without any effect on the coercivity [81].

Comparably low values of $(BH)_{max}$ achieved for Mn-based $L1_0$ ordered thin films are related to relatively small coercivity provided by these materials.

4.4   Spintronic Applications of $L1_0$ Ordered Magnetic Thin Films

Magnetic tunnel junctions (MTJs) are nanostructured spintronics devices attracting considerable interest due to their applications in modern technologies such as magnetic sensors and magnetic random access memories (MRAM). The variety of physical phenomena, which govern the functionality of these magnetoresistive devices, makes MTJs also very attractive from the fundamental physics point of view. Even after many years of research in the field, this stimulates intensive activities in the experimental and theoretical investigations of electronic, magnetic and transport properties of MTJs [113, 118].

In the simplest case, MTJs consist of two ferromagnetic (FM) layers separated by a thin insulating tunnel barrier. One of the FM layers has its magnetization fixed and acts as a reference layer. The magnetic moment of the second FM layer is free to switch between two opposite states – parallel and antiparallel with respect to the fixed layer. These two magnetization orientations correspond to the two different resistance states of the MTJ structure – the states with low and high resistance. In 1996, Slonczewski and Berger predicted theoretically that a spin-polarized current could reversibly change and even switch the magnetization direction of a FM layer in multilayer spintronic structures. This effect was called spin-transfer torque (STT) [1, 93]. Shortly after, this effect was experimentally confirmed giving rise to the era of the STT-based technologies.

It turned out that a spin-polarized dc current can cause precession of the magnetization in a free magnetic layer with frequencies lying in the microwave range. Thus, a rather small dc current (above the excitation threshold of magnetization dynamics) passing through an MTJ structure can emit microwaves in the range of 10 GHz. Devices based on this phenomena are called spin torque nano-oscillators (STNO). They can be used in a wide range of applications including neuromorphic computing, microwave generation, detection, and modulation [4, 50]. The output power of these devices is in the range 10–100 nW and can be enhanced using arrays of STNOs [20, 31, 75] or relying on their structural optimization [24, 91].

First STT experiments were performed for spintronic multilayer structures having thin metal spacers (typically made of Cu), and therefore exhibiting the giant magnetoresistance (GMR) effect [33, 36, 39, 84]. Ultra-fast thermally-assisted magnetization switching by STT was experimentally studied in [9]. It was also shown that a substantial improvement of the device performance can be obtained for spintronic structures utilizing in-plane magnetized free layer and out-of-plane magnetized pinned magnetic layer possessing a strong PMA [54]. This motivated the use of $L1_0$ ordered magnetic alloys as a component of STNO architectures [60]. According to the estimations made in [60], the use of $L1_0$ ordered magnetic materials can significantly increase the output power efficiency compared to STNOs based on GMR stacks with in-plane magnetization [36, 39, 75, 77, 83]. Experiments performed in [34] also proved a substantial reduction of the switching current in an MTJ with PMA (Fig. 4.7).

Major improvement in the performance of STNOs came with the use of oxide barriers based on amorphous AlO_x- or crystalline MgO ultrathin films [10, 26, 107]. Compared to the GMR-based counterparts, these MTJs rely on tunneling magnetoresistance (TMR) effect and possess substantially large resistances (~ 1–10 kΩ), large microwave output powers (~ 1 μW), large output dc signals (~ 1 mV) if they are used as microwave signal rectifiers, as well as better compatibility with the standard complementary-symmetry/metal-oxide semiconductor (CMOS) fabrication technology [28, 33, 90].

Fig. 4.7 Switching current density at 10 ns write pulse width for both switching directions in an MTJ as a function of the free layer thickness showing a substantial reduction of the current in the samples containing layers with perpendicular magnetic anisotropy. (Reproduced from [34] with the permission of AIP Publishing)

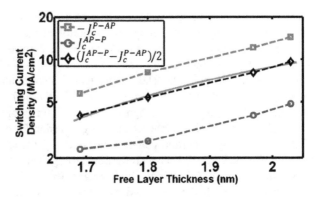

There are three important parameters characterizing the performance of MTJs in reading, writing, and storing data [87]:

1. Tunneling magnetoresistance ratio:

$$TMR = \frac{\Delta R}{R} = \frac{R_{AP} - R_P}{R_P},\qquad(4.1)$$

where R_{AP} and R_P—resistance of the element when magnetization of the free and pinned layers are antiparallel or parallel, respectively.

2. Thermal stability factor:

$$\Delta = \frac{E_b}{k_B T},\qquad(4.2)$$

where E_b— energy barrier preventing the thermally induced magnetization reversal; k_B—Boltzmann constant; T— absolute temperature.

3. Critical switching current:

$$I_c \tilde{\alpha} M_S H_K,\qquad(4.3)$$

Where α—damping constant; H_K— anisotropy field.

To use MTJs in non-volatile spintronic applications, a high TMR ratio (in the order of 100%), a high thermal stability factor (depends on the junction capacitance but generally higher than 60), and a low I_c (smaller than the driving current of the cell transistor) are required [100]. All three conditions were successfully achieved for CoFeB-based MTJs, which are implemented in commercial STT-MRAM magnetic memory devices (Everspin[1]). However further improvement is needed for increase of memory density and the energy consumption of these devices.

To fulfill the first condition, a perfect interface between FM and barrier layers is required. It means that the lattice mismatch between the FM and barrier layers has to be as minimal as possible. The second requirement could be satisfied only if the energy of the magnetic anisotropy is sufficiently high. Therefore, FM material should reveal high value of magnetocrystalline anisotropy due to the tendency to scale down the MTJ volume, since energy barrier E_b proportional to K_u and V. Finally, the third condition could be achieved by reducing M_s or α. However, it should be noted that K_u is also proportional to M_s. Thus, the fabrication of materials with small M_s while keeping a high magnetocrystalline anisotropy is one of the material science challenges posed by modern spintronic technologies. Moreover, these materials should reveal a low damping constant.

Relying on the STT effect, spin-torque diodes (SD) were proposed, which are able to rectify the microwave current in an MTJ [57, 78, 79, 102]. It was shown that if the free MTJ's layer reveals perpendicular magnetization, high frequency SD can be obtained without an external magnetic field application [13, 66, 80, 101]. The fact is that the frequency of ferromagnetic resonance (f_{FMR}) is proportional to the effective anisotropy field H_k^{eff}, and inversely proportional to the magnetic damping constant α. Thus, using PMA materials with strong magnetic anisotropy and small M_s and α is an efficient approach to create high performance spin-torque diodes operating in a wide range of frequencies [65]. Another distinct but promising regime of the SD operation is a non-resonance regime, first observed in experiments with thermally-activated "non-adiabatic stochastic resonance" [7] and then theoretically described in [76, 77, 79]. These diodes work as broadband low-frequency non-resonant microwave detectors and could be used for energy harvesting applications. These devices possess no resonance frequency, and, therefore, could harvest energy from the entire low-frequency region of the microwave spectrum (below the threshold frequency). The energy conversion rate may reach about 3.5%, that is sufficient for applications [76]. Recently, a possibility of the microwave energy harvesting in a non-resonant SD possessing a strong PMA in the free layer has been experimentally demonstrated (Fig. 4.8) [15].

$L1_0$ ordered hard magnetic alloys offer several advantages for MTJ structures. Strong perpendicular anisotropy of these films could allow the realization of the so-called perpendicular MTJs (p-MTJs) with high thermal stability. Moreover, Mn-based alloys reveal small damping constant. Therefore, their use could lead to the reduction of the switching current while keeping high thermal stability. Furthermore, alloys with a strong PMA allow to boost the performance of spin-torque diodes operating in a broad frequency range and promising for energy harvesting applications. Hence, $L1_0$ chemically ordered thin films such as FePt, CoPt, FePd [29, 43, 61, 74, 110, 115], MnAl [71, 73, 82, 108], MnGa [44, 63, 104, 106] attract broad research interest.

---

[1]https://www.everspin.com/spin-transfer-torque-mram-products

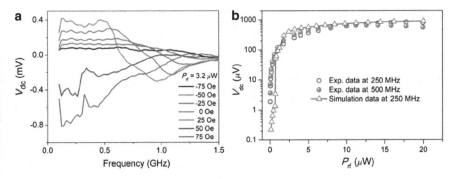

Fig. 4.8 Frequency (left figure) and power (right figure) dependence of the output dc voltage of a SD operating in the energy harvesting mode. (**a**) The rectification curve under various in-plane magnetic fields at rf input of 3.2 μW. (**b**) A comparison of experimental and simulated data of the rectification voltage as a function of input power. (Reprinted with permission from [15] Copyright (2020) by the American Physical Society)

Inami et al. investigated magnetic and magneto-transport properties of perpendicularly magnetized MTJ stack of Ta(5 nm)/CoPt(30 nm)/MgO(2.5 nm)/FePt (4 nm)/Pt(10 nm)/Cr(40 nm)/MgO(001) [29]. CoPt and FePt layers were sputtered at 675 K and 775 K, respectively, to obtain the ordered $L1_0$ structure with perpendicular magnetization. The thinnest FePt layer, which reveals PMA was found to be 4 nm thick. A clear switching between parallel and antiparallel magnetization of FM layers was found in the field of ~2 kOe. However, rather low TMR ratio of about 6% was obtained at room temperature. Yang et al. prepared perpendicular MTJs stacks of FePt(5 nm)/MgO(3 nm)/FePt(20 nm) by molecular beam epitaxy and received TMR ratio of 21% and 53% at 300 K and 10 K, respectively [112].

A relatively large TMR exceeding 100% at room temperature was obtained by Yoshikawa et al. using polycrystalline FePt/MgO/Fe/FePt films prepared by magnetron sputtering on thermally oxidized Si(001) substrates [115]. Cross-sectional high resolution transmission electron microscopy (TEM) image of the stacks show that a large lattice mismatch of 9.3% between the MgO barrier and $L1_0$-FePt electrode prevents the epitaxial growth of the layers in the [001] direction. It was shown that introducing an additional ultrathin Fe layer between the barrier and the bottom ferromagnetic layers allows to relax the lattice mismatch and to form a perpendicularly magnetized epitaxial heterostructure. Large TMR values could be obtained due to a strong PMA of the Fe(001)/$L1_0$-FePt(001) bottom electrode and high quality MgO(001) barrier. Furthermore, the optimization of the heat treatment conditions were done to realize a sufficient difference between the coercivities of the top and bottom $L1_0$-FePt ferromagnetic layers.

$L1_0$ ordered CoPt alloys were also used as FM electrodes in perpendicular MJTs not only because of their large K_u but also due to the lower – compared to FePt – saturation magnetization of 800 emu/cm^3. This is beneficial for lowering of the current density needed for spin-transfer switching of the magnetization. The TMR ratio of Ta(5 nm)/CoPt(30 nm)/MgO(3 nm)/CoPt(20 nm)/Pt(3 nm)/Cr(40 nm)/MgO (001) MTJs was investigated by Kim et al. [35]. Ar gas pressure during magnetron

Fig. 4.9 Theoretical TMR
calculations for MnAl/
MgO-based MTJs.
(Reproduced from [116]
with the permission of AIP
Publishing)

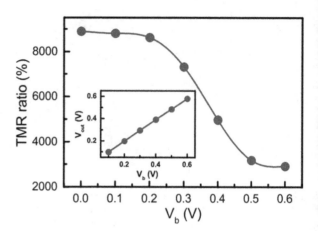

sputtering of the bottom CoPt layer as well as the substrate temperature were
optimized and found to provide best structural and magnetic properties with values
of 0.65 Pa and 873 K, respectively. The top CoPt electrode was sputtered at higher
Ar pressure of 1.2 Pa, providing higher coercivity of the layer. The resulting TMR
ratios were found to be 6% and 13% at 300 K and 10 K, respectively. Low TMR
values were explained by a big mismatch between the MgO and CoPt lattices leading
to the appearance of interfacial dislocations and reduction of the spin polarization
due to the unfavorable location of the $\Delta 1$ band.

According to theoretical calculations, high spin polarization in MgO-based MTJs
and a rather low magnetic damping are expected in ordered $L1_0$-MnAl alloys
[82]. Zhang et al. predicted theoretically a remarkably large TMR ratio of up to
2000% for MnAl/MgO-based MTJs using first principle calculations (Fig. 4.9)
[116]. The TMR effect in MnAl-based MTJs was measured by Saruyama et al.
using CoFe as a free layer and underlayer between MnAl and the MgO barrier
[85]. The MnAl alloy revealed a small Ms of 530 emu/cm^3 and high Ku of 1×10^7
erg/cm^3. Still, a TMR ratio of only 2.1% was reported. Recently, MnAl-based MTJs
with a thin Co_2MnSi intermediate layer were produced by molecular-beam epitaxy
revealing 10% TMR at 5 K [56]. The low TMR ratio was explained by the existence
of defects in the MnAl layer near the MgO barrier.

The situation is similar when MnGa thin films are used for MTJs. Theoretical
calculations suggest excellent TMR up to 600% [42]. However, high TMR has not
been measured experimentally yet. For instance Miyazaki's et al. reported MnGa/
MgO/CoFe MTJs with TMR ratio of about 23% at 10 K [42]. Recent studies report
p-MTJ MnGa/Co/MgO/CoFeB structures with TMR ratio of up to 40% at room
temperature. Introducing a Co layer in the stack allows to increase the TMR and to
decrease the resistance area and magnetization values (Fig. 4.10) [51, 97].

The main reason why MnAl- and MnGa-based MTJs do not exhibit high TMR
ratio is attributed to the large lattice mismatch (up to 7%) with the MgO barrier layer
[71, 97]. The $L1_0$-MnAl phase has lattice constants of $a = 0.394$ nm and
$c = 0.356$ nm [71]. The $L1_0$-phase of MnGa has lattice constants of $a = 0.388$ nm
and $c = 0.364$ nm [97]. At the same time, MgO has lattice parameter $a = 0.4212$ nm.

Fig. 4.10 Co thickness (t_{Co}) dependence of the MR ratio (**a**), resistance area (R_PA) (**b**), and remanent magnetization (M_r) of MnGa/Co bilayers (**c**). (Reproduced from [51] with the permission of AIP Publishing)

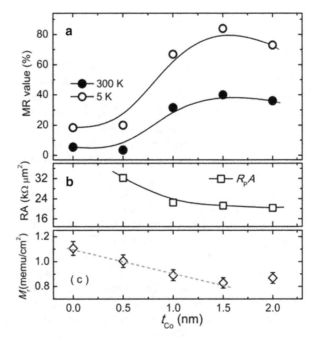

It is expected that the reduction of the lattice mismatch between the Mn-based hard magnetic alloy layers and MgO will lead to the enhancement of the TMR ratio. One of the prospective approaches for lowering of the lattice mismatch is to introduce an additional intermediate layer with appropriate lattice parameters such as Cr, CrRu, CoFe, Co$_2$MnSi, CoGa [42, 51, 56, 64, 71] or introduction of an additional layer outside of MTJ to form a strained MnGa layer with an increased lattice parameter [97]. Another way to improve bonding between the MgO and MnAl(MnGa) layers is to tune the lattice parameter of Mn-based alloys by incorporation of light elements (H, C, N, B) using heat treatment in an appropriate atmosphere or upon deposition.

Oogane et al. demonstrated that Cr and CrRu are promising intermediate layers to reduce the lattice mismatch between MnAl and MgO layers [71, 73, 105]. For instance, using Cr$_{10}$Ru$_{90}$ as a buffer layer allowed to achieve a small value of M_s (500 emu/cm^3) with a roughness of about 0.34 nm while keeping a high value of K_u (13 × 10^6 erg/cm^3) [73]. It has been shown that Co substitution into the MnAl layer is an effective approach to reduce the saturation magnetization (from 560 to 350 emu/cm^3) and surface roughness down to 0.2 nm [105]. The impact of non-magnetic layers (e.g. Cr or CrRu) on the TMR performance is yet to be investigated.

Inserting additional FM buffer layers is considered as another promising route towards high TMR. Using two FM layers, one of which has strong PMA and the second one improves bonding with the MgO barrier, could allow achieving high TMR, high thermal stability and low power consumption [23]. One of the benefits of such a layer stack is that two FM layers in MTJs could be exchange coupled (ferromagnetic or antiferromagnetic coupling) forming a so-called synthetic antiferromagnet. Switching of one layer could lead to the switching of the second one. This approach was tested on $L1_0$ ordered MnGa-based MTJs. For example, MnGa/Co exchange coupled systems were investigated, and MnGa/Co/MgO/CoFeB MTJs with TMR of about 40% at room temperature has been demonstrated [51]. Introducing Co_2MnSi between the MnGa and MgO layers was reported by S. Mao and TMR of 40% at 10 K was achieved in the MnGa/Co_2MnSi/MgO/Co_2MnSi/MnGa MTJs. Both the top and bottom MnGa electrodes show a pronounced PMA and a strong antiferromagnetic coupling in the MnGa/Co_2MnSi bi-layer [56]. A high TMR ratio of 27% was achieved in MnGa(25 nm)/Co(0.8 nm)/MgO(2.3 nm)/Fe(5 nm)/Pd (2 nm) p-MTJ stacks grown using molecular beam epitaxy on GaAs substrates [117]. A thin Co layer was used to improve the interface quality between the MnGa and MgO and reduce the lattice mismatch. A strong antiferromagnetic coupling was observed between the MnGa and Co layers.

Very recently, Suzuki et al. realized ultrathin MnGa layer (1 nm) in combination with CoGa layer (30 nm) and applied it as a free layer in the MTJ structure. CoFeB was used as a fixed layer. To increase the TMR ratio, the interface between the MnGa and MgO was modified by adding several monolayers of Mn. The samples with inserted Mn layers revealed an increased TMR ratio compared to the reference samples without the Mn insertion (TMR increase from 1.7% to 19%). Furthermore, the samples did not show an increase of the M_s. It was observed that MnGa and Mn have ferromagnetic coupling at the interface [98]. Because of the large bulk magnetic anisotropy of the bottom FM layer, such systems can be a promising solution for further downscaling of MTJ cells below 10 nm.

4.5   Conclusions

Magnetic thin films are key elements of state-of-the-art functional devices and have potential to impact prospective electronic and spintronic technologies. Among broad variety of magnetic alloys, $L1_0$ ordered thin films typically attract much attention primarily due to their excellent structural and magnetic properties but also due to their high thermodynamic stability. The key feature of most $L1_0$ nanosized structures is their strong magnetocrystalline magnetic anisotropy. This feature makes $L1_0$ ordered alloys attractive for permanent magnet applications. Furthermore, $L1_0$ thin films show extremely high tunability of such magnetic properties like saturation magnetization, coercivity and magnetic damping constant. Small values of both saturation magnetization and damping constant of Mn-based $L1_0$ thin films make them promising candidates for MRAM and STNO applications.

Extensive experimental and theoretical efforts are put into the exploration of the potential of thin films and heterostructures for permanent magnet applications. Magnetic properties of nanocomposites can be controlled by tuning the deposition parameters, alloy composition, or applying post-processing treatment. The exchange spring coupling (ESC) effect is a promising approach to improve the PMs performance of thin film magnets. ESC based on $L1_0$-FePt hard magnetic alloys exhibit $(BH)_{max}$ values higher than 50 MGOe, which is in the range of modern RE containing PMs. The current state of research with $L1_0$ ordered MnAl- and MnGa-hard magnetic alloys reveals rather moderate PM performance primarily due to their small coercive field. However, these alloys do not contain Pt, which make them attractive for further exploratory research aiming to increase the energy product. Several problems have to be solved to improve the PM performance of Mn-based alloys. In MnAl alloys, Mn often replaces Al atoms leading to a Mn-Mn antiferromagnetic coupling and the following magnetization decrease. Another aspect, where much work is still need to be done is related to the investigation of Mn alloys-based ESC effect. The ESC depends on many factors including magnetic properties of hard and soft phases, their thickness, geometry of hard-soft system, interface quality (structural and magnetic), surface/interface roughness. All these aspects have to be understood for Mn-based alloys to make a final conclusion on their application potential for PM applications.

There are several issues that have to be solved to create MRAM devices based on $L1_0$ materials with increased thermal stability, memory density and lower power consumption compared to current CoFeB based devices. Another promising spintronic application direction of $L1_0$ materials includes the development and practical use of microwave and sub-terahertz signal oscillators, detectors, spectrum analyzers, energy harvesting devices. In such devices the use of $L1_0$ materials instead of conventional magnetic alloys potentially allows one to substantially reduce the devices' working current density, increase its operation frequency and/or ac/dc energy conversion rate.

References

1. Berger L (1996) Emission of spin waves by a magnetic multilayer traversed by a current. Phys Rev B 54:9353
2. Bhatti S, Sbiaa R, Hirohata A et al (2017) Spintronics based random access memory: a review. Mater Today 20:530–548
3. Calvayrac F, Bajorek A, Randrianantoandro N (2018) Mechanical alloying and theoretical studies of MnAl (C) magnets. J Magn Magn Mater 462:96–104
4. Choi HS, Kang SY, Cho SJ et al (2014) Spin nano–oscillator–based wireless communication. Sci Rep 4:1–7
5. Chen J, Sun C, Chow GM (2008) A review of L10 FePt films for high-density magnetic recording. Int J Prod Dev 5:238
6. Crisan O, Vasiliu F, Crisan AD (2019) Structure and magnetic properties of highly coercive L10 nanocomposite FeMnPt thin films. Mater Charact 152:245–252
7. Cheng X, Boone CT, Zhu J et al (2010) Nonadiabatic stochastic resonance of a nanomagnet excited by spin torque. Phys Rev Lett 105:047202
8. Dang R, Ma X, Liu J et al (2017) The fabrication and characterization of MnAl/FeCo composite thin films with enhanced maximum energy products. Mater Lett 197:8–11
9. Devolder T, Tulapurkar A, Yagami K et al (2005) Ultra-fast magnetization reversal in magnetic nano-pillars by spin-polarized current. J Magn Magn Mater 286:77–82

10. Diao Z, Apalkov D, Pakala M et al (2005) Spin transfer switching and spin polarization in magnetic tunnel junctions with MgO and AlO x barriers. Appl Phys Lett 87:232502

11. Endoh T, Honjo H (2018) A recent progress of spintronics devices for integrated circuit applications. J Low Power Electron Appl 8:44

12. Fang B, Carpentieri M, Hao X et al (2016) Giant spin-torque diode sensitivity in the absence of bias magnetic field. Nat Commun 7:1–7

13. Fang H, Kontos S, Ångström J et al (2016) Directly obtained τ-phase MnAl, a high performance magnetic material for permanent magnets. J Solid State Chem 237:300–306

14. Fang H, Cedervall J, Casado FJM et al (2017) Insights into formation and stability of τ-MnAlZx (Z= C and B). J Alloys Compd 692:198–203

15. Fang B, Carpentieri M, Louis S et al (2019) Experimental demonstration of spintronic broadband microwave detectors and their capability for powering nanodevices. Phys Rev Appl 11:014022

16. Feng JN, Liu W, Gong WJ et al (2017) Magnetic properties and coercivity of MnGa films deposited on different substrates. J Mat Sci Tech 33:291–294

17. Fischer GA, Rudee ML (2000) Effect of magnetic annealing on the τ-phase of MnAl thin films. J Magn Magn Mater 213:335–339

18. Fullerton EE, Jiang JS, Grimsditch M et al (1998) Exchange-spring behavior in epitaxial hard/soft magnetic bilayers. Phys Rev B 58:12193

19. Gavrea R, Hirian R, Mican S et al (2017) Structural, electronic and magnetic properties of the Mn54− xAl46Tix (x= 2; 4) alloys. Intermetallic 82:101–106

20. Georges B, Grollier J, Cros V et al (2008) Impact of the electrical connection of spin transfer nano-oscillators on their synchronization: an analytical study. Appl Phys Lett 92:232504

21. Giannopoulos G, Reichel L, Markou A et al (2015) Optimization of L10 FePt/Fe45Co55 thin films for rare earth free permanent magnet applications. J Appl Phys 117:223909

22. Herzer G (1990) Grain size dependence of coercivity and permeability in nanocrystalline ferromagnets. IEEE Trans Magn 26:1397–1402

23. Hou-Fang L, Ali SS, Xiu-Feng H (2014) Perpendicular magnetic tunnel junction and its application in magnetic random access memory. Chin Phys B 23:077501

24. Houssameddine D, Florez SH, Katine JA et al (2008) Spin transfer induced coherent microwave emission with large power from nanoscale MgO tunnel junctions. Appl Phys Lett 93:022505

25. Huang EY, Kryder MH (2015) Fabrication of MnAl thin films with perpendicular anisotropy on Si substrates. J Appl Phys 117:17E314

26. Huai Y, Pakala M, Diao Z et al (2006) Spin-transfer switching in MgO magnetic tunnel junction nanostructures. J Magn Magn Mater 304:88–92

27. Huang EY, Kryder HM (2015) L1$_0$-ordered MnAl thin films with high perpendicular magnetic anisotropy using tin underlayers on Si substrates. Paper presented at the INTERMAG 2015 – IEEE international magnetics conference, Beijing, China, 11–15 May 2015

28. Ikeda S, Hayakawa J, Ashizawa Y et al (2008) Tunnel magnetoresistance of 604% at 300 K by suppression of Ta diffusion in Co Fe B/ Mg O/ Co Fe B pseudo-spin-valves annealed at high temperature. Appl Phys Lett 93:082508

29. Inami N, Kim G, Hiratsuka T et al (2010) Structural, magnetic, and magnetotransport properties of FePt/MgO/CoPt perpendicularly magnetized tunnel junctions. J Phys Conf Ser 200:052008

30. Jian H, Skokov KP, Gutfleisch O (2015) Microstructure and magnetic properties of Mn–Al–C alloy powders prepared by ball milling. J Alloys Compd 622:524–528

31. Kaka S, Pufall MR, Rippard WH et al (2005) Mutual phase-locking of microwave spin torque nano-oscillators. Nature 437:389–392

32. Kamino K, Kawaguchi T, Nagakura M (1996) Magnetic properties of MnAl system alloys. IEEE Trans Magn 2:506–510

33. Katine JA, Fullerton EE (2008) Device implications of spin-transfer torques. J Magn Magn Mater 320:1217–1226

34. Khalili Amiri P, Zeng ZM, Langer J et al (2011) Switching current reduction using perpendicular anisotropy in CoFeB–MgO magnetic tunnel junctions. Appl Phys Lett 98:112507
35. Kim G, Sakuraba Y, Oogane M et al (2008) Tunneling magnetoresistance of magnetic tunnel junctions using perpendicular magnetization L 1 0-Co Pt electrodes. Appl Phys Lett 92:172502
36. Kiselev SI, Sankey JC, Krivorotov IN et al (2003) Microwave oscillations of a nanomagnet driven by a spin-polarized current. Nature 425:380–383
37. Kneller EF, Hawig R (1991) The exchange-spring magnet: a new material principle for permanent magnets. IEEE Trans Magn 27:3588–3560
38. Kontos S, Fang H, Li J et al (2019) Measured and calculated properties of B-doped τ-phase MnAl–A rare earth free permanent magnet. J Magn Magn Mater 474:591–598
39. Krivorotov IN, Emley NC, Sankey JC et al (2005) Time-domain measurements of nanomagnet dynamics driven by spin-transfer torques. Science 307:228–231
40. Krone P, Makarov D, Schrefl T et al (2010) Exchange coupled composite bit patterned media. Appl Phys Lett 97:082501
41. Krone P, Makarov D, Schrefl T et al (2011) Correlation of magnetic anisotropy distributions in layered exchange coupled composite bit patterned media. J Appl Phys 109:103901
42. Kubota T, Ma Q, Mizukami S et al (2012) Dependence of tunnel magnetoresistance effect on Fe thickness of perpendicularly magnetized L10-Mn62Ga38/Fe/MgO/CoFe junctions. Appl Phys Exp 5:043003
43. Kugler Z, Grote JP, Drewello V et al (2012) Co/Pt multilayer-based magnetic tunnel junctions with perpendicular magnetic anisotropy. J Appl Phys 111:07C703
44. Kurt H, Rode K, Venkatesan M et al (2011) High spin polarization in epitaxial films of ferrimagnetic Mn 3 Ga. Phys Rev B 83:020405
45. Laenens B, Almeida FM, Planckaert N et al (2009) Interplay between structural and magnetic properties of L 1 0-FePt (001) thin films directly grown on MgO (001). J Appl Phys 105:073913
46. Liu X, Morisako A (2008) Soft magnetic properties of FeCo films with high saturation magnetization. J Appl Phys 103:07E726
47. Liu Y, George TA, Skomski R et al (2011) Aligned and exchange-coupled FePt-based films. Appl Phys Lett 99:172504
48. Liu Y, Sellmyer DJ (2016) Exchange-coupling behavior in nanostructured FePt/Fe bilayer films. AIP Adv 6:056010
49. Lodder JC, Nguyen LT (2005) FePt thin films: fundamentals and applications. In: Encyclopedia of materials: science and technology. Elsevier, pp 1–10
50. Louis S, Sulymenko O, Tiberkevich V et al (2018) Ultra-fast wide band spectrum analyzer based on a rapidly tuned spin-torque nano-oscillator. Appl Phys Lett 113:112401
51. Ma QL, Kubota T, Mizukami S et al (2012) Magnetoresistance effect in L 10-MnGa/MgO/CoFeB perpendicular magnetic tunnel junctions with Co interlayer. Appl Phys Lett 101:032402
52. Makarov D, Lee J, Brombacher C et al (2010) Perpendicular FePt-based exchange-coupled composite media. Appl Phys Lett 96:062501
53. Makarov A, Windbacher T, Sverdlov V et al (2016) CMOS-compatible spintronic devices: a review. Semicond Sci Technol 31:113006
54. Mangin S, Ravelosona D, Katine JA et al (2006) Current-induced magnetization reversal in nanopillars with perpendicular anisotropy. Nat Mater 5:210–215
55. Mao S, Lu J, Zhao X et al (2017) MnGa-based fully perpendicular magnetic tunnel junctions with ultrathin Co 2 MnSi interlayers. Sci Rep 7:1–7
56. Mao S, Lu J, Wang H et al (2019) Observation of tunneling magnetoresistance effect in L10-MnAl/MgO/Co2MnSi/MnAl perpendicular magnetic tunnel junctions. J Phys D Appl Phys 52:405002
57. Matsumoto R, Chanthbouala A, Grollier J et al (2011) Spin-torque diode measurements of MgO-based magnetic tunnel junctions with asymmetric electrodes. Appl Phys Exp 4:063001

58. McCallum RW, Lewis LH, Skomski R et al (2014) Practical aspects of modern and future permanent magnets. Annu Rev Mater Res 44:451–477
59. Mican S, Benea D, Hirian R et al (2016) M. Structural, electronic and magnetic properties of the Mn50Al46Ni4 alloy. J Magn Magn Mater 401:841–847
60. Mitani S (2011) Spin-transfer magnetization switching in ordered alloy-based nanopillar devices. J Phys D Appl Phys 44:384003
61. Moriyama T, Mitani S, Seki T et al (2004) Magnetic tunnel junctions with L1 0-ordered FePt alloy electrodes. J Appl Phys 95:6789–6791
62. Mohapatra J, Liu JP (2018) Rare-Earth-free permanent magnets: the past and future. In: Handbook of magnetic materials, vol 27. Elsevier, pp 1–57
63. Mizukami S, Kubota T, Wu F et al (2012) Composition dependence of magnetic properties in perpendicularly magnetized epitaxial thin films of Mn-Ga alloys. Phys Rev B 85:014416
64. Mizukami S, Kubota T, Iihama S et al (2014) Magnetization dynamics for L 10 MnGa/Fe exchange coupled bilayers. J Appl Phys 115:17C119
65. Mizukami S, Sugihara A, Iihama S et al (2016) Laser-induced THz magnetization precession for a tetragonal Heusler-like nearly compensated ferrimagnet. Appl Phys Lett 108:012404
66. Naganuma H, Kim G, Kawada Y et al (2015) Electrical detection of millimeter-waves by magnetic tunnel junctions using perpendicular magnetized L 10-FePd free layer. Nano Lett 15:623–628
67. Navío C, Villanueva M, Céspedes E et al (2018) Ultrathin films of L10-MnAl on GaAs (001): a hard magnetic MnAl layer onto a soft Mn-Ga-As-Al interface. APL Mater 6:101109
68. Nie SH, Zhu LJ, Lu J et al (2013) Perpendicularly magnetized τ-MnAl (001) thin films epitaxied on GaAs. Appl Phys Lett 102:152405
69. Nozaki T, Yamamoto T, Miwa S et al (2019) Recent Progress in the voltage-controlled magnetic anisotropy effect and the challenges faced in developing voltage-torque MRAM. Micromachines 10:327
70. Ono A, Suzuki KZ, Ranjbar R et al (2017) Ultrathin films of polycrystalline MnGa alloy with perpendicular magnetic anisotropy. Appl Phys Exp 10:023005
71. Oogane M, Watanabe K, Saruyama H et al (2017) L10-ordered MnAl thin films with high perpendicular magnetic anisotropy. Japn J Appl Phys 56:0802A2
72. Øygarden V, Rial J, Bollero A et al (2019) Phase-pure τ-MnAlC produced by mechanical alloying and a one-step annealing route. J Alloys Compd 779:776–783
73. Park JH, Hong YK, Bae S et al (2010) Saturation magnetization and crystalline anisotropy calculations for MnAl permanent magnet. J Appl Phys 107:09A731
74. Peng SZ, Zhang Y, Wang M X et al (2005) in Wiley Encyclopedia of electrical and electronics engineering, pp 1–16
75. Prokopenko O, Bankowski E, Meitzler T et al (2011) A spin-torque nano-oscillator as a microwave signal source. IEEE Magn Lett 2:3000104–3000104
76. Prokopenko OV, Krivorotov IN, Bankowski E et al (2012) Spin-torque microwave detector with out-of-plane precessing magnetic moment. J Appl Phys 111:123904
77. Prokopenko OV, Krivorotov IN, Bankowski EN et al (2013) Hysteresis regime in the operation of a dual-free-layer spin-torque nano-oscillator with out-of-plane counter-precessing magnetic moments. J Appl Phys 114:173904
78. Prokopenko OV, Krivorotov IN, Meitzler TJ et al (2013) Spin-torque microwave detectors. Magnonics. Springer, Berlin/Heidelberg, pp 143–161
79. Prokopenko OV, Slavin AN (2015) Microwave detectors based on the spin-torque diode effect. Low Temp Phys 41:353–360
80. Rippard WH, Deac AM, Pufall MR et al (2010) Spin-transfer dynamics in spin valves with out-of-plane magnetized CoNi free layers. Phys Rev B 81:014426
81. Sabet S, Moradabadi A, Gorji S et al (2019) Correlation of interface structure with magnetic exchange in a hard/soft magnetic model nanostructure. Phys Rev Appl 11:054078
82. Sakuma A (1994) Electronic structure and magnetocrystalline anisotropy energy of MnAl. J Phys Soc Jpn 63:1422–1428

83. Sanches F, Tiberkevich V, Guslienko KY et al (2014) Current-driven gyrotropic mode of a magnetic vortex as a nonisochronous auto-oscillator. Phys Rev B 89:140410

84. Sankey JC, Cui YT, Sun JZ et al (2008) Measurement of the spin-transfer-torque vector in magnetic tunnel junctions. Nat Phys 4:67–71

85. Saruyama H, Oogane M, Kurimoto Y et al (2013) Fabrication of L10-ordered MnAl films for observation of tunnel magnetoresistance effect. Jpn J Appl Phys 52:063003

86. Sato T, Ohsuna T, Kaneko Y (2016) Enhanced saturation magnetization in perpendicular L 10–MnAl films upon low substitution of Mn by 3 d transition metals. J Appl Phys 120:243903

87. Sato H, Ikeda S, Ohno H (2017) Magnetic tunnel junctions with perpendicular easy axis at junction diameter of less than 20 nm. Jpn J Appl Phys 56:0802A6

88. Sato K, Takahashi Y, Makuta H et al (2018) Dependence of thickness of Fe buffer layer on magnetic properties for Mn2. 6Ga thin films. Trans Magn Soc Jpn 2:48–51

89. Sbiaa R, Piramanayagam SN (2017) Recent developments in spin transfer torque MRAM. Phys Stat Sol Rapid Res Lett 11:1700163

90. Silva TJ, Rippard WH (2008) Developments in nano-oscillators based upon spin-transfer point-contact devices. J Magn Magn Mater 320:1260–1271

91. Sim CH, Moneck M, Liew T et al (2012) Frequency-tunable perpendicular spin torque oscillator. J Appl Phys 111:07C914

92. Skomski R, Manchanda P, Kumar P et al (2013) Predicting the future of permanent-magnet materials. IEEE Trans Magn 49:3215–3220

93. Slonczewski JC (1996) Current-driven excitation of magnetic multilayers. J Magn Magn Mater 159:L1

94. Su KP, Chen XX, Wang HO et al (2016) Effect of milling on the structure and magnetic properties in Mn54Al46 flakes prepared by surfactant-assisted ball milling. Mater Charact 114:263–266

95. Suess D, Schrefl T, Fähler S et al (2005) Exchange spring media for perpendicular recording. Appl Phys Lett 87:012504

96. Süss D (2006) Multilayer exchange spring media for magnetic recording. Appl Phys Lett 89:113105

97. Suzuki KZ, Ranjbar R, Okabayashi J et al (2016) Perpendicular magnetic tunnel junction with a strained Mn-based nanolayer. Sci Rep 6:30249

98. Suzuki KZ, Miura Y, Ranjbar R et al (2018) Perpendicular magnetic tunnel junctions with Mn-modified ultrathin MnGa layer. Appl Phys Lett 112:062402

99. Takahashi Y, Makuta H, Shima T et al (2017) Fabrication and magnetic properties of L10-MnGa highly oriented thin films. Trans Magn Soc Jpn Special Issues 1:30–33

100. Takemura R, Kawahara T, Miura K et al (2010) A 32-Mb SPRAM with 2T1R memory cell, localized bi-directional write driver and1'/0'dual-array equalized reference scheme. IEEE J Solid State Circuits 45(4):869–879

101. Taniguchi T, Arai H, Tsunegi S et al (2013) Critical field of spin torque oscillator with perpendicularly magnetized free layer. Appl Phys Exp 6:123003

102. Tulapurkar AA, Suzuki Y, Fukushima A et al (2005) Spin-torque diode effect in magnetic tunnel junctions. Nature 438:339–342

103. Vashisht G, Goyal R, Bala M et al (2018) Studies of exchange coupling in FeCo/L1 0-FePt bilayer thin films. IEEE Trans Magn 55:1–5

104. Wang K, Lu E, Knepper JW et al (2011) Structural controlled magnetic anisotropy in Heusler L 1 0– MnGa epitaxial thin films. Appl Phys Lett 98:162507

105. Watanabe K, Oogane M, Ando Y (2017) Cobalt substituted L10-MnAl thin films with large perpendicular magnetic anisotropy. Jpn J Appl Phys 56:0802B1

106. Weller D, Parker G, Mosendz O, Lyberatos A et al (2016) FePt heat assisted magnetic recording media. J Vacuum Sci Technol B Nanotechnol Microelectron 34:060801

107. Wen ZC, Wei HX, Han XF (2007) Patterned nanoring magnetic tunnel junctions. Appl Phys Lett 91:122511

108. Wurmehl S, Kandpal HC, Fecher GH et al (2006) Valence electron rules for prediction of half-metallic compensated-ferrimagnetic behaviour of Heusler compounds with complete spin polarization. J Phys Condens Matter 18:6171

109. Xiang Z, Song Y, Deng B et al (2019) Enhanced formation and improved thermal stability of ferromagnetic τ phase in nanocrystalline Mn55Al45 alloys by Co addition. J Alloys Compd 783:416–422

110. Yakushiji K, Saruya T, Kubota H et al (2010) Ultrathin co/Pt and co/Pd superlattice films for MgO-based perpendicular magnetic tunnel junctions. Appl Phys Lett 97:232508

111. Yan ZC, Huang Y, Zhang Y et al (2005) Magnetic and structural properties of MnAl/ag granular thin films with L10 structure. Scr Mater 53:463–468

112. Yang G, Li DL, Wang SG et al (2015) Effect of interfacial structures on spin dependent tunneling in epitaxial L 10-FePt/MgO/FePt perpendicular magnetic tunnel junctions. J Appl Phys 117:083904

113. Yang MS, Fang L, Chi YQ (2018) Dependence of switching process on the perpendicular magnetic anisotropy constant in P-MTJ. Chin Phys B 27:098504

114. You J, Guo Y (2018) Plasma enhanced atomic layer deposition of Co thin film on τ-MnAl for effective magnetic exchange coupling and enhanced energy products. J Alloys Compd 758:116–121

115. Yoshikawa M, Kitagawa E, Nagase et al (2008) Tunnel magnetoresistance over 100% in MgO-based magnetic tunnel junction films with perpendicular magnetic L10-FePt electrodes. IEEE Trans Magn 44:2573–2576

116. Zhang X, Tao LL, Zhang J et al (2017) First-principles study of MnAl for its application in MgO-based perpendicular magnetic tunnel junctions. Appl Phys Lett 110:252403

117. Zhao XP, Lu J, Mao SW et al (2017) L10-MnGa based magnetic tunnel junction for high magnetic field sensor. J Phys D Appl Phys 50:285002

118. Zhong H, Wen Y, Zhao Y et al (2019) Ten states of nonvolatile memory through engineering ferromagnetic remanent magnetization. Adv Funct Mater 29:1806460

119. Zhu L, Nie S, Meng K et al (2012) Multifunctional L10-Mn1. 5Ga films with ultrahigh coercivity, Giant perpendicular Magnetocrystalline anisotropy and large magnetic energy product. Adv Mater 24:4547–4551

Chapter 5
Effect of Au Layers on $A1 \rightarrow L1_0$ Phase Transition and Magnetic Properties of FePt Thin Films

Pavlo Makushko, Mark Shamis, Tetiana Verbytska, Sergii Sidorenko, and Iurii Makogon

Abstract Complex effect of annealing conditions (temperature, duration, atmosphere (vacuum, hydrogen)), thickness of intermediate Au layer, mechanical residual stresses and strains on the formation of a hard-magnetic $L1_0$ phase as well as structural and magnetic properties of [FePt(15 nm)/Au(7.5–30 nm)/FePt(15 nm)]$_n$ ($n = 1$; 2) layer stacks deposited by magnetron sputtering onto SiO$_2$(100 nm)/Si (001) substrates was investigated.

It was shown that stress/strain level in FePt layers strongly depends on the Au interlayer thickness, leading to different onset temperatures of $L1_0$-FePt phase formation processes. Compressive strain occurring in the as-deposited FePt layers promotes lowering of the $L1_0$ ordering temperature.

Applying annealing in hydrogen, the $A1$ into $L1_0$-FePt phase transition in the FePt/Au/FePt stacks starts at 500 °C regardless of the intermediate Au layer thickness. The Au and FePt axial (111) textures were observed in the films annealed in this atmosphere. Higher values of coercivity can be obtained at lower temperatures compared to annealing in vacuum.

Keywords Ordered $L1_0$-FePt phase · Mechanical stress · Annealing · Coercivity · Vacuum

5.1 Introduction

The equiatomic FePt alloy with ordered $L1_0$ structure is a promising material for application as an ultrahigh density magnetic recording medium due to its excellent magnetic properties. In particular this phase reveals high magnetocrystalline

P. Makushko (✉) · M. Shamis · T. Verbytska · S. Sidorenko · I. Makogon
Metals Physics Department, National Technical University of Ukraine "Igor Sikorsky Kyiv Polytechnic Institute", Kyiv, Ukraine
e-mail: makushko@kpm.kpi.ua

© Springer Nature B.V. 2020
A. Kaidatzis et al. (eds.), *Modern Magnetic and Spintronic Materials*, NATO
Science for Peace and Security Series B: Physics and Biophysics,
https://doi.org/10.1007/978-94-024-2034-0_5

anisotropy of up to 7×10^7 erg/cm^3 [27, 40]. Wu et al. showed prototype which combines FePt-C as recording medium and heat assisted magnetic recording technique, areal recording density of 1 Tbit/inch^2 was obtained [42]. The extension of areal density to 5 Tbit/inch^2 using this material seems to be feasible [14, 19]. For this purpose, highly ordered $L1_0$-FePt films with easy magnetization axis oriented perpendicular to the film plane, revealing perpendicular magnetic anisotropy, are required.

Direct deposition of FePt alloy at room temperature results in formation of disordered soft-magnetic $A1$-FePt phase. The ordered $L1_0$-FePt phase could be formed during deposition at elevated temperatures or at subsequent post-annealing process [9, 43, 52]. Numerous approaches, such as different deposition techniques, preparation of layer stacks, third element alloying have been investigated to obtain (001) textured highly ordered $L1_0$-FePt films.

It was found that stress/strain state has significant influence on the $L1_0$ ordering processes and oriented growth of FePt thin films [16, 17, 20, 21, 34, 41]. Internal stress in FePt films can be tuned by variation of deposition and post-annealing parameters, substrate type, film thickness, and alloying. In Refs. [1, 15, 39, 49] it was shown that tensile stress induced by rapid thermal annealing, due to the difference in light absorption coefficients of FePt and substrate promotes $L1_0$ ordering and (001) preferred orientation of FePt grains.

However, external stress can either assist or hinder the $L1_0$ phase transformation of the FePt alloy. In this regard, it was shown that large compressive stress leads to increase of the energy barrier of ordering reaction. In turn, strong tensile stress blocks formation of $L1_0$ phase by preventing densification – a vital process prior to ordering [41, 49]. Furthermore, the $L1_0$ ordering processes can be accelerated by interface reactions. For instance, Li et al. showed that direct interface reaction between FePt film and Si substrate resulting in formation of PtSi interface layer introduces dynamic stress and thus enhances diffusion mobility of Fe and Pt atoms [25]. Intrinsic strain originated from lattice mismatch with substrate material can also be used for tuning magnetic properties of FePt thin films. Futamoto et al. showed that epitaxial FePt film on MgO substrate (-10.3% lattice mismatch) possesses stronger perpendicular magnetization properties, compared to films grown on substrates with lower lattice mismatch [10]. These effects can be tuned by variation of FePt film thickness.

Alloying of FePt films with additional elements was found to be an effective approach for tuning its magnetic properties. Great attention is attracted to doping of FePt thin films with elements revealing low surface energy such as Au [8, 33, 37, 47], Ag [18, 32, 36, 51], and Cu [2, 28, 40, 50]. For instance, additional stresses and interface energy induced by addition of Au as well as its diffusion along FePt grain boundaries could result in achievement of perpendicular magnetic anisotropy, enhanced coercivity and decreased ordering temperature [33, 37, 47]. Moreover, thermally-induced diffusion of Au atoms along FePt grain boundaries could hinder FePt grain growth [8]. Alloying of FePt with 9 at. % of Cu also was found to be effective for enhancing of ordering processes in FePt thin films. Cu atoms replace Fe in Fe-rich planes of $L1_0$ structure and enhance both Fe atoms mobility and tetragonality of ordered structure. Thus, high perpendicular anisotropy of FePtCu films can be obtained [28, 35, 41, 50]. Positive impact of alloying by such elements as Cr [50], Ni [38], B [12], W, Ti and Ru [26] has been proved as well.

Annealing atmosphere could provide significant effects on the $L1_0$ ordering. Yamauchi et al. [43] showed that annealing in hydrogen atmosphere accelerates the structural changes rate during $A1{\rightarrow}L1_0$ ordering processes in AuCu nanoparticles by 100 times compared to vacuum annealing. In Ref. [34] it was shown that oxygen content complicates the mutual diffusion between Fe and Pt layers during annealing in inert gases and vacuum. Contrastingly, the penetration of hydrogen atoms into the crystalline lattice enhances Fe and Pt atoms mobility and thus could decrease the temperature of $L1_0$-FePt phase formation [24, 31].

The aim of present study is to investigate the effect of annealing conditions (temperature, duration, atmosphere (vacuum, hydrogen)) on residual stress/strain state and $L1_0$-FePt ordered phase formation in [FePt(15 nm)/Au(7.5–30 nm)/FePt (15 nm)]$_n$ (n = 1; 2) layer stacks.

The effect of Au interlayer thickness on stress/strain state, structural and magnetic properties of FePt/Au/FePt layer stacks will be discussed in section A of experimental results. Also, the effect of total thickness of the stack will be presented in section B, for this purpose, [FePt/Au/FePt]$_2$ stacks were prepared. The effectiveness of hydrogen atmosphere as an approach for enhancing ordering processes in FePt/ Au/FePt stacks will be investigated in section C.

5.2   Experimental

The FePt-based stacks with various stacking were deposited by DC magnetron sputtering using individual Fe, Pt and Au targets on thermally oxidized (100 nm-thick SiO_2) single crystalline Si(001) substrate at room temperature. The base pressure of sputtering chamber before deposition was 1×10^{-7} mbar and Ar work pressure was kept as 3.5×10^{-3} mbar. The sputtering rates were monitored *in-situ* by quartz microbalance approach. The thicknesses of as-deposited layers were confirmed by Rutherford backscattering technique. The sputtering rates of Fe and Pt were adjusted to obtain layers of equiatomic $Fe_{50}Pt_{50}$ alloy.

Post-annealing of the stacks was carried out in vacuum of $\sim 10^{-3}$ Pa vacuum and hydrogen atmosphere ($\sim$100 kPa) in temperature range from 300 to 900 °C. The annealing duration at chosen temperature varied from 30 s to 1 h, while heating rates were adjusted to 5 and 1 °C/s, respectively for vacuum and hydrogen annealing.

The structural properties of film samples were characterized by X-ray diffraction (XRD) (ULTIMA IV Rigaku diffractometer) using Θ-2Θ geometry, in Cu k_α radiation. In-plane residual stresses in FePt layers were evaluated by X-ray tensometry $\sin^2\psi$ approach [7, 49] using (111) FePt peak. Here we should note, that in case of as-deposited FePt/Au/FePt stack the (111) FePt peak on XRD pattern is a sum of reflections from two separated FePt layers, and thus obtained value of stress should be understood as an average value. The residual strain $\varepsilon_\perp$ along the surface normal was extracted using the following equation

$$\varepsilon_\perp = \frac{d_{\psi=0} - d_{bulk}}{d_{bulk}}, \tag{5.1}$$

where d_{bulk} is the stress-free lattice spacing of the $A1$ or $L1_0$ bulk phases (2.235 Å and 2.19 Å, respectively [3, 29]).

The $L1_0$ ordering degree in FePt alloy was evaluated using two approaches: estimation of intensity ratio between superlattice (001) and fundamental (002) peaks [6, 45], and estimation of c/a tetragonality degree of corresponding structure [46]. A fraction of FePt grains with c axis being oriented perpendicular to the film plane is evaluated from the $I(001)/I(111)$ intensity ratio.

The average crystallites size was evaluated by Scherrer equation

$$D_{av} = \frac{0.9 \cdot \lambda}{\beta \cdot \cos \Theta} \tag{5.2}$$

using (111) peak of FePt, where λ is the X-ray wavelength, β is full width at intensity half maximum of the peak, and Θ is the peak angle position.

Magnetic properties were evaluated by superconductive quantum interference device – vibrating sample magnetometer (SQUID-VSM) applying in-plane and out-of-plane magnetic fields at room temperature. Polar magneto-optical Kerr effect (MOKE) magnetometry was also used.

Change of surface morphology with annealing temperature and corresponding variation of surface roughness were evaluated using atomic force microscopy (AFM) technique.

5.3 Results and Discussion

5.3.1 Effect of Au Interlayer Thickness on the Phase Transformations and Magnetic Properties of FePt/Au/ FePt Layer Stacks during Annealing in Vacuum

The FePt(15 nm)/Au(x nm)/FePt(15 nm) layer stacks with Au layer thickness x = 7.5; 15; 20; 30 nm were prepared in order to investigate the influence of Au interlayer thickness on $L1_0$ ordering processes in FePt layers. The FePt(30 nm) thin film was used as reference (XRD of this sample is not shown).

The XRD patterns of the as-deposited and post-annealed FePt/Au/FePt stacks are summarized in Fig. 5.1. All films after deposition reveal diffraction peaks from disordered $A1$-FePt phase and Au. Signal from the Si substrate also could be seen on diffraction spectras. Reflections from Au become stronger with increase of Au interlayer thickness. Post-annealing at temperatures below 500 °C did not lead to noticeable structural changes (not shown).

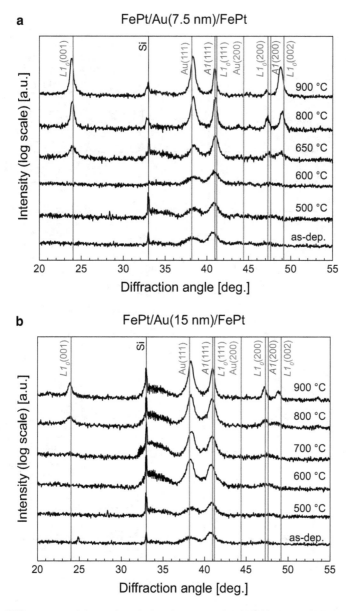

Fig. 5.1 XRD patterns of the as-deposited and post-annealed FePt/Au(x nm)/FePt (x = 7.5 (**a**); 15 (**b**); 20 (**c**); 30 nm (**d**)) stacks

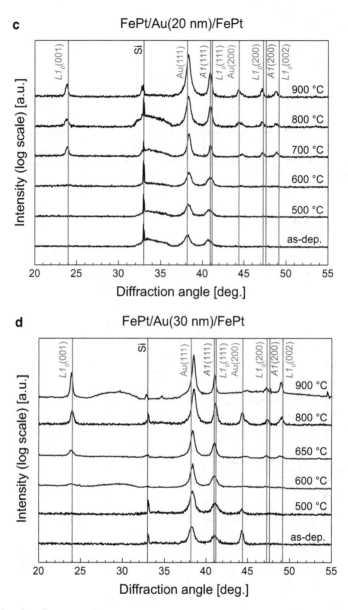

Fig. 5.1 (continued)

In the films with 7.5 nm- and 15 nm-thick intermediate Au layers, $L1_0$-FePt ordered phase starts to be formed during annealing in vacuum at temperatures of 650 °C and 700 °C, correspondingly (Fig. 5.1a, b), these temperatures are higher compared to the ordering temperature obtained for FePt(30 nm) reference sample (600 °C). Formation of ordered phase is ascertained by a change of FePt crystal structure from face-centered cubic (*fcc*) to face-centered tetragonal (*fct*). Formation of *fct* ordered structure is indicated by the appearance of superstructure (001) FePt peak and splitting of (200) FePt peak into (200) and (002) reflections. Increase of Au interlayer thickness to 20 nm and 30 nm results in decrease of ordering temperature down to 600 °C.

Annealing at temperatures above 700 °C are accompanied by pronounced ordering processes in the FePt layers, as reflected by the splitting of the (200) peak and raised intensity of the superstructure (001) peak (Fig. 5.1). However, no preferred (001) orientation of $L1_0$-FePt grains was obtained.

We suggest that this difference in $L1_0$ ordered phase formation temperatures is associated with initial stress/strain state of the films. Additional stress in FePt layers, induced by introduction of Au interlayer, originate from difference of thermal expansion coefficients (10.5×10^{-6} K^{-1} [34] for FePt and 14×10^{-6} K^{-1} for Au) and from strain energy of FePt/Au interface [8]. The effect of Au interlayer thickness on the stress/strain state of FePt layers in as-deposited and corresponding change of ordering temperature are shown in Fig. 5.2. The stress/strain state varies with the Au interlayer thickness and at some critical thickness the process of stress/strain relaxation occurs and may be explained inelastic deformation [13]. This is reflected by a stress sign change from compressive to tensile in the sample with 15 nm-thick Au interlayer, causing reduction of the strain (Fig. 5.2b). According to the Le Chatelier-Braun principle, this effect should lead to an increase in ordering temperature as was observed experimentally (Fig. 5.2a). Further increase of Au

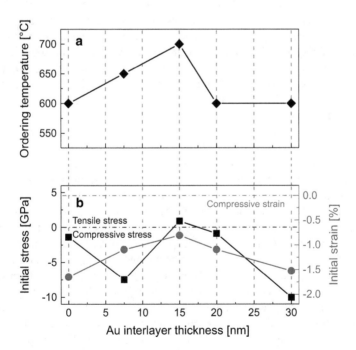

Fig. 5.2 Ordering temperature (**a**) and initial stress/strain state in FePt layers (**b**) as a function of Au interlayer thickness

interlayer thickness up to 30 nm is accompanied by a rise of initial compressive stress and strain and, correspondingly, decrease in ordering temperature. Thus, a clear correspondence between residual stress/strain of the FePt/Au/FePt as-deposited stacks and the onset temperature of chemical ordering was observed.

Introduction of Au interlayer changes the effect of initial stress on ordering in FePt layers. The effect of stress on ordering in FePt thin films as well as its variation with annealing temperature was investigated by Hsiao et al. [15, 17]. It was shown that low level initial tensile stress is more suitable condition for $L1_0$ ordering compared to compressive stress. During annealing accumulation of tensile strain takes place, which is a result of densification process due to defect elimination and grain growth. And at some critical level this strain is released via the nucleation of $L1_0$-FePt phase [15]. Introduction of Au interlayer complicates evolution of stress/strain state during annealing. Additional stresses are introduced into the FePt layers from difference in thermal expansion coefficients of FePt and Au, lattice mismatch, and diffusion of Au along FePt grain boundaries.

The residual stress/strain state varies further during thermal annealing due to the difference in thermal expansion coefficients of FePt, Au and the Si substrate. While the stress state does not change significantly with annealing temperature, a strong variation of the residual strain which becomes tensile upon $L1_0$ phase formation was observed (Fig. 5.3b). The latter fact is caused by the decrease of lattice volume during $A1$-FePt (V = 58.8 Å^3) → $L1_0$-FePt (V = 54.7 Å^3) phase transition [3, 29].

The ordering degree of the $L1_0$-FePt phase is expressed by the integral intensity ratio between the (001) superstructure and (002) fundamental peaks. This ratio increases with annealing temperature and gets more pronounced in the films with thicker Au interlayers (Fig. 5.4a). From the intensity ratio of the (001) and (111) reflections it is clear that Au interlayer introduction suppresses the $L1_0$-FePt grain growth with c axis being perpendicular to the film plane (Fig. 5.4b). This effect is caused by preferred (111) orientation of Au grains, distribution of which along FePt grain boundaries is expected [8, 48].

The tetragonal distortions of fcc $A1$-FePt phase crystal lattice during thermally-activated $L1_0$ ordering processes are affected by addition of Au interlayer. As it can be seen from Fig. 5.5, the a lattice constant increases with rise of annealing temperature while decrease of both c constant and c/a ratio is observed, being indicative of increase in the tetragonality and ordering degree of the $L1_0$-FePt phase (Fig. 5.5b, c). However, the tetragonality degree was decreased by the addition of Au. This effect was also observed by Platt et al. [33], where larger values of a lattice constant in the FePt film alloyed by Au were obtained, and in Ref. [8] it was suggested that excess of Au in Fe/Pt/Au multilayer stack suppresses ordering during high temperature heat treatment.

The structural properties of FePt/Au/FePt stacks have significant effect on their magnetic behavior, which was investigated by SQUID-VSM magnetometry. The dependence of out-of-plane coercivity (OOP) on the Au interlayer thickness after annealing at 900 °C is shown in Fig. 5.6c. The coercivity of the film with 7.5 nm-thick Au layer increases from 10.2 kOe to 16.55 kOe as annealing temperature rises from 650 to 900 °C. An increase of annealing duration up to 1 h is required to obtain similar values of coercivity in reference FePt(30 nm) films. OOP coercivity of stacks after annealing at 900 °C rises with Au interlayer thickness (Fig. 5.6b). Maximum

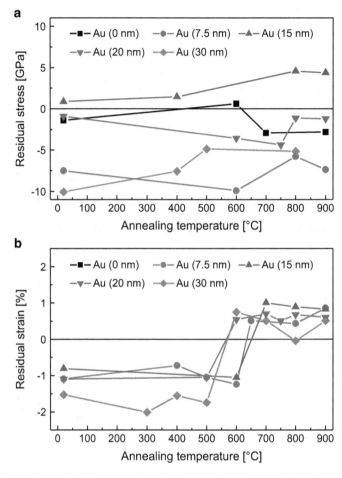

Fig. 5.3 Dependence of residual stress (**a**) and strain (**b**) in FePt layers of FePt/Au/FePt stacks on the annealing temperature

coercivity of 27.3 kOe was observed for film with 30 nm-thick Au interlayer. This increase in coercivity can be associated with two factors: (i) increased fraction of $L1_0$ ordered grains in FePt/Au/FePt stacks and (ii) diffusion of Au along FePt grain boundaries, resulting in decrease of magnetic exchange interaction between them. Diffusion of Au, which has (111) texture, along FePt grain boundaries suppressed formation of (001) texture in $L1_0$-FePt grains. The film reveals magnetically isotropic behavior since no preferred orientation of easy magnetization axis of hard magnetic grains was obtained.

The similar effect of Au in FePt films was observed in [4], where annealing of Au/FePt/Au stack resulted in formation of magnetically isolated ordered FePt particles which were distributed in Au matrix. As a result, the coercivity was doubled compared to pure FePt film. In Refs. [8, 11] it was theoretically and experimentally

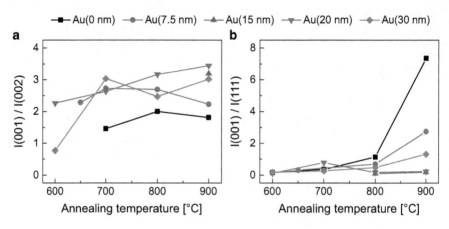

Fig. 5.4 Dependence of I(001)/I(002) (**a**) and I(001)/I(111) (**b**) intensity ratios on annealing temperature

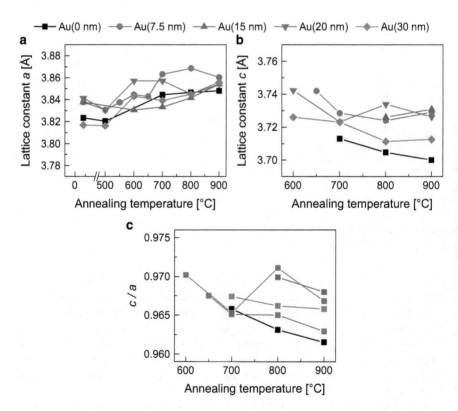

Fig. 5.5 Change of a (**a**) and c (**b**) lattice constants as well as c/a ratio (**c**) of FePt phase with annealing temperature

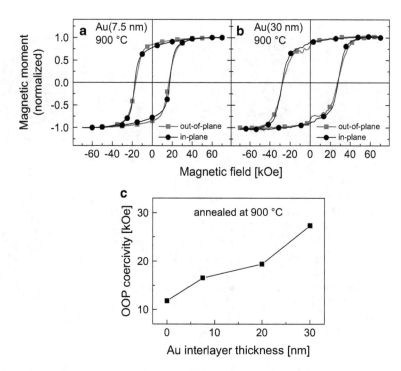

Fig. 5.6 Normalized M-H hysteresis loops of FePt/Au/FePt stacks annealed in vacuum at temperature of 900 °C (**a, b**) and effect of Au layer thickness on coercivity of the stacks (**c**)

confirmed, that increase of Au concentration along FePt grain boundaries also rises strain anisotropy and pins the magnetic domain walls, which also has an impact on magnetic behavior of the stack. Such an effect was also shown in Fe/Pt/Au multilayers [47], where the intrinsic stress, stored in as-deposited films, was found to be an additional driving force for the rearrangement of Fe and Pt atoms during $A1$ to $L1_0$ transformation. However, annealing at high temperatures and increased amount of Au resulted in ordered phase fraction.

All as-deposited thin films have smooth surface with average roughness of about 0.25 nm. Annealing in vacuum at high temperatures leads to increase in surface roughness (Fig. 5.7). Greater average surface roughness of films with 7.5 nm- and 30 nm-thick Au interlayer may be related to greater amount of (001) oriented grains of $L1_0$-FePt phase (Fig. 5.4b).

In summary, addition of Au interlayer influences the initial stress/strain state in FePt layers, which affects the ordering temperature. High initial compressive stress in FePt layers promotes ordering during subsequent annealing. Diffusion of Au along FePt grain boundaries forms non-magnetic walls and promotes (111) preferred orientation.

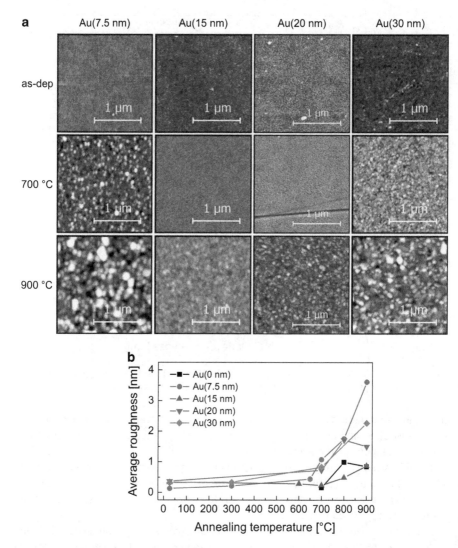

Fig. 5.7 AFM images of FePt/Au/FePt stacks (**a**) and corresponding change of average surface roughness with annealing temperature (**b**)

5.3.2 Influence of Number of Au Interlayers on A1-FePt→L1$_0$-FePt Phase Transition and Magnetic Properties in [FePt/Au/FePt]$_2$ Layer Stack

The [FePt(15 nm)/Au(7.5 nm)/FePt(15 nm)]$_2$ layer stack was prepared in order to investigate the influence of number of Au interlayers and total thickness of stack on the stress state, ordering processes and magnetic properties of FePt thin films by comparison with results obtained for FePt and FePt/Au(7.5 nm)/FePt films in section A.

XRD patterns of the as-deposited and post-annealed [FePt/Au/FePt]$_2$ stacks are shown in Fig. 5.8a. The as-deposited thin film reveals reflections from chemically disordered $A1$-FePt phase, Au and silicon substrate. Formation of $L1_0$-FePt phase in the [FePt/Au/FePt]$_2$ stack was observed after annealing in vacuum at 800 °C, which is for 150 °C higher compared to the film with single Au(7.5 nm) interlayer (corresponding results were discussed in previous section). This increase of onset temperature of $L1_0$ ordered phase formation caused by lower level of initial compressive stress in FePt layers of the stack due to the larger stack thickness (Fig. 5.8b). According to the principles, described in previous section, this lowering of initial compressive stress results in less intensive ordering in FePt layers, and thus, higher onset temperature of $L1_0$ ordered phase formation was obtained. Investigation of magnetic properties with MOKE approach indicates corresponding decrease of

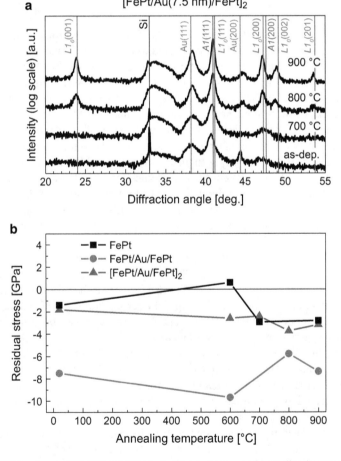

Fig. 5.8 XRD patterns of [FePt/Au/FePt]$_2$ stack after deposition and annealing in vacuum (**a**) and change of residual stress in FePt layers with annealing temperature (**b**)

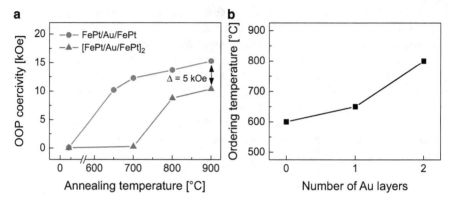

Fig. 5.9 Change of coercivity of [FePt/Au/FePt]$_n$ (n = 1; 2) layer stacks with annealing temperature (**a**) and effect of Au layers number on ordering temperature (**b**)

coercivity (Fig. 5.9a). After annealing at 900 °C the stack with two Au interlayers possess coercivity for 5 kOe lower compared to the film with one Au layer.

Increase of $I(001)/I(002)$ ratio as temperature rise from 800 to 900 °C and higher values of c/a ratio (Fig. 5.10a, b) indicate that in [FePt/Au/FePt]$_2$ stack ordering processes proceed at higher temperatures and are less intensive compared to reference samples. These results contradict those, obtained by Minh et al., according to which, an increase of the thickness of FePt film leads to an acceleration of ordering, since barrier energy for ordered phase nucleation is lower in thicker films [30]. From the plot on Fig. 5.10c it is clearly seen, that (001) axis of $L1_0$-FePt phase in [FePt/Au/FePt]$_2$ stack is in-plane oriented. Several studies [5, 22, 23] indicate that at some critical thickness, which is around 30 nm, (001) orientation of $L1_0$-FePt films decreases and (111) texture become dominant in order to minimize surface energy. This process is accompanied by a decrease in perpendicular magnetic anisotropy and exchange coupling of FePt grains.

Thermally-activated ordering processes occurring during annealing have effect on the surface morphology of [FePt/Au/FePt]$_2$ stacks which is reflected in corresponding changes of roughness (Fig. 5.13a, b). Surface roughness as well as grain size of FePt phase, obtained using Scherrer equation, are almost constant at annealing temperatures below 650 °C. The ordering processes occurring at higher temperatures were accompanied by an increase of both surface roughness as well as FePt crystallites size (Fig. 5.13c). Surface roughness of [FePt/Au/FePt]$_2$ stack after high temperature annealing is much lower compared to the film with one Au interlayer. This is due to less pronounced ordering processes in FePt layers, which are accompanied with grain growth of ordered phase and development of their orientation (Fig. 5.11).

From these facts it can concluded that an increase of both Au(7.5 nm) layers number and total thickness of the stack results in a slow-down of $L1_0$ ordering processes in FePt/Au/FePt stacks, which is expressed in a higher ordering

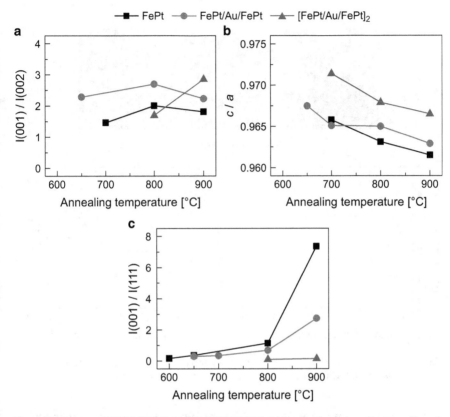

Fig. 5.10 Change of I(001)/I(002) (**a**) and I(001)/I(111) (**b**) integral intensity ratios as well as c/a ratio (**c**) of $L1_0$-FePt phase in [FePt/Au/FePt]n (n = 1; 2) stacks with annealing temperature

temperature and decrease in coercivity. In this case, a significant decrease in surface roughness was observed.

5.3.3   Effect of Annealing in Hydrogen Atmosphere on Phase Transformations and Magnetic Properties of FePt/Au/FePt Stacks

Possibility to enhance $L1_0$ ordering processes by introduction of hydrogen in heat treatment atmosphere was shown in [43]. It was suggested that hydrogen exposure treatment can be applied to enhance functional properties of $L1_0$ structured functional materials, such as thin films of ordered FePt and FePd alloys. In order to investigate that the FePt/Au/FePt stacks were subjected to heat treatment in hydrogen atmosphere.

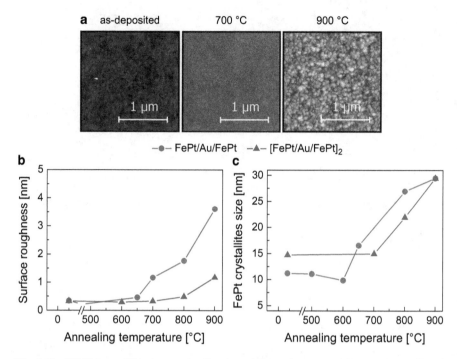

Fig. 5.11 AFM images (**a**) and corresponding dependences of average surface roughness (**b**) and FePt grain size (**c**) in [FePt/Au/FePt]$_n$ (n = 1; 2) stacks on annealing temperature

The XRD patterns of FePt/Au(7.5; 30 nm)/FePt stacks after annealing in hydrogen atmosphere are shown in Fig. 5.12. Low intensity peaks corresponding to ordered $L1_0$-FePt phase were observed after annealing at 500 °C regardless of the Au interlayer thickness. This decrease in ordering temperature, comparing to annealing in vacuum, is attributed to penetration of hydrogen atoms into FePt crystalline lattice, increase of its parameters, induction of additional compressive stresses and decrease in Fe-Pt binding energy.

Further increase in annealing temperature results in a rise of $L1_0$ ordered phase amount, which is indicated by increased intensity of (001) and (002) reflections. Additional reflections appear angles of $2\Theta = 32$, 34.5, and 36.7° on XRD patterns from film sample with thicker Au(30 nm) interlayer intensity of which increases with annealing temperature. It is assumed that these peaks can belong to $AuH_{0.35}$ hydride formed as a result of hydrogen atoms introduction into the film. Although $AuH_{0.35}$ is unstable at room temperature, this compound can be observed under nonequilibrium conditions of nanosized films.

The ordering processes in FePt layers occur with decrease in constant c and c/a ratio, although constant a barely changes with annealing temperature (Fig. 5.13). Increased c/a ratio after annealing at 800 °C most probably caused by lattice

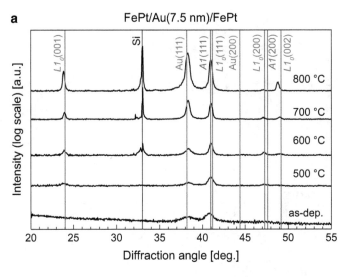

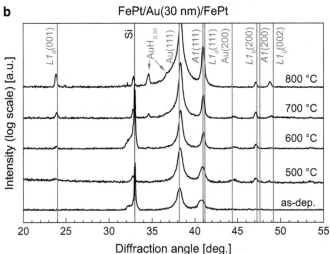

Fig. 5.12 XRD patterns of the FePt/Au(x nm)/FePt (x = 7.5 (**a**); 30 nm (**b**)) stacks annealed in hydrogen

distortions due to intensive penetration of hydrogen atoms into octahedral and tetrahedral voids of the FePt crystalline lattice.

From *I(001)/I(002)* ratio it can be reasonably assumed, that ordering degree of FePt phase increases with annealing temperature. This process is more pronounced in film with 30 nm-thick Au interlayer (Fig. 5.14a). However, ordering degree was found to be higher in stack with Au(7.5 nm) interlayer, since thicker Au layer can restrain growth of *fct* FePt grains during annealing [47]. From the *I(001)/I(111)* ratio (Fig. 5.14b) it can be concluded, that (001) axis of $L1_0$ ordered structure is in-plane

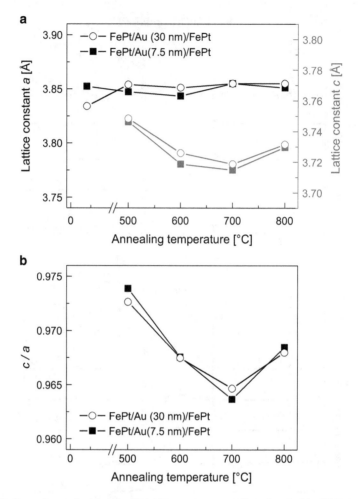

Fig. 5.13 Dependence of a (**a**) and c (**b**) lattice constants as well as c/a ratio (**c**) of FePt in FePt/Au (7.5; 30 nm)/FePt stacks on the temperature of annealing in hydrogen

oriented. Enhanced diffusion rate of Fe, Pt and Au and additional distortions of crystalline lattice contributed to the formation of (111) texture of both FePt and Au, since it is more energetically preferable.

The similar effect of hydrogen content in annealing atmosphere was observed in Ref. [36], where it was shown that oxygen/hydrogen content plays important role in control of crystallographic orientation of $L1_0$ phase. The introduction of hydrogen atoms into FePt film at annealing in forming gas (Ar + 3% H) suppresses oriented grain growth in direction perpendicular to the film plane.

Introduction of hydrogen into annealing atmosphere promotes ordering in FePt films by means of several factors: (i) Chemical interaction of penetrated into film hydrogen atoms with Fe and Pt atoms weakens Fe-Pt atomic bonds and, thus,

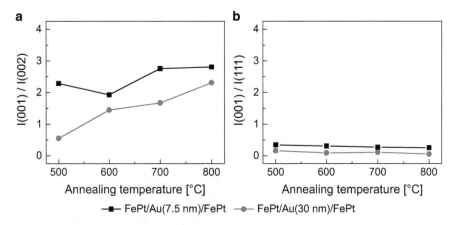

Fig. 5.14 Effect of temperature of annealing in hydrogen on I(001)/I(002) (**a**) and I(001)/I(111) (**b**) intensity ratios of FePt peaks in the FePt/Au(7.5; 30 nm)/FePt stacks

enhances their diffusion rate [43]. (ii) Hydrogen reduces oxygen content in the film and atmosphere of heat treatment. Oxygen in FePt films reacts with Fe atoms resulting in formation of stable Fe-O oxides, usually at grain boundaries. This prohibits nucleation of ordered phase crystallites and, thus, increases $L1_0$ ordering temperature. Leistner et al. showed that annealing in hydrogen reduces oxygen content in electrodeposited FePt film by three times, resulting in increased magnetization [24]. (iii) Penetration of hydrogen atoms into voids of crystalline lattice of the film promote its additional distortions and provides additional strain energy, which also enhances rearrangement and atomic mobility of Fe and Pt atoms during ordering.

The corresponding enhancement of magnetic properties by hydrogen annealing was observed by SQUID magnetometry measurements (Fig. 5.15). Annealing at 500 °C resulted in $L1_0$ ordered phase formation with coercivities of 10.3 kOe and 18.6 kOe in thin films with 7.5 nm- and 30 nm-thick Au interlayer, respectively. The maximum coercivity (27.3 kOe) of the stack with an intermediate 30 nm-thick Au layer was observed after annealing in hydrogen at 700 °C, and the same coercivity value can be reached by vacuum annealing at 900 °C.

In summary, annealing in hydrogen atmosphere affects ordering in FePt/Au/FePt stacks. Ordering in investigated stacks starts at 500 °C regardless of the Au layer thickness. This decrease, comparing to annealing in vacuum (discussed in sections A – C) can be attributed to additional distortions, caused by penetration of hydrogen atoms into FePt lattice. Higher coercivities can be achieved at lower temperatures comparing with annealing in vacuum.

Fig. 5.15 The change in coercivity of FePt/Au(7.5; 30 nm)/FePt stacks with temperature of annealing in hydrogen

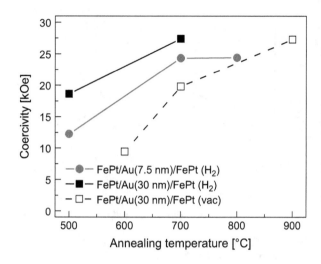

5.4 Conclusions

The $L1_0$ ordering processes in [FePt/Au/FePt]$_n$ (n = 1; 2) stacks were investigated. It was shown that initial stress/strain state in FePt layers, which can be tailored by Au interlayer thickness, has a significant influence on the $L1_0$ ordering processes. Large compressive stress in the as-deposited FePt layers induced by thicker Au interlayer promotes $L1_0$ ordered phase formation. Diffusion of Au atoms along FePt grain boundaries reduces magnetic exchange interaction between magnetic grains, and, as a result, enhances coercivity of thin films. Thus, coercivity of 27.3 kOe can be reached in films with 30 nm-thick interlayer. Simultaneously, introduction of Au interlayer suppresses the formation of (001) texture in FePt layers.

It was shown that increase of both number of Au layers and total thickness of [FePt/Au/FePt]$_n$ (n = 1; 2) stack reduces the level of compressive stress in FePt layers, lowering $L1_0$ ordering and leading to increase of ordering temperature up to 800 °C.

The tetragonal lattice distortions that occur during ordering process most intensively develop during first 30 s of annealing. Further annealing is accompanied with increase in $L1_0$ ordered phase amount.

It was shown that introduction of hydrogen into annealing atmosphere is an effective approach to lower onset temperature of $L1_0$ phase formation and enhance its coercivity. Coercivity of 27.3 kOe in FePt/Au(30 nm)/FePt stack was obtained after annealing in hydrogen at lower temperature compared to the annealing in vacuum.

Acknowledgements This work was financially supported by the German Academic Exchange Service (DAAD) in the frame of Leonard Euler scholarship program (Grant ID 55576194, Grant ID 57042790, Grant ID 57198300, Grant ID 57291435).

References

1. Albrecht M, Brombacher C (2013) Rapid thermal annealing of Fe P t thin films. Phys Stat Sol (a) 210:1272–1281
2. Brombacher C, Schletter H, Daniel M et al (2012) FePtCu alloy thin films: morphology, L10 chemical ordering, and perpendicular magnetic anisotropy. J Appl Phys 112:073912
3. PDF Card № 00-029-0717, Cabri LJ, Feather CE (1975) Platinum-iron alloys: a nomenclature based on a study of natural and synthetic alloys. Can Mineral Can Mineral 13:117
4. Chen SK, Yuan FT, Liao WM et al (2006) Magnetic properties and microstructure study of high coercivity Au/FePt/Au trilayer thin films. J Magn Magn Mater 303:e251–e254
5. Chen SC, Sun TH, Chang CL et al (2011) Film thickness dependence of microstructures and magnetic properties in single-layered FePt films by in-situ annealing. Thin Solid Films 519:6964–6968
6. Christodoulides JA, Farber P, Dannl M et al (2001) Magnetic, structural and microstructural properties of FePt/M (M= C, BN) granular films. IEEE Trans Magn 37:1292–1294
7. Cullity BD (1956) Elements of X-ray diffraction. Addison-Wesley Publishing
8. Feng C, Zhan Q, Li B et al (2008) Magnetic properties and microstructure of FePt/Au multilayers with high perpendicular magnetocrystalline anisotropy. Appl Phys Lett 93:152513
9. Futamoto M, Nakamura M, Ohtake M et al (2016) Growth of L 10-ordered crystal in FePt and FePd thin films on MgO (001) substrate. AIP Adv 6:085302
10. Futamoto M, Nakamura M, Shimizu T et al (2018) Influence of stress and strain on $L1_0$-ordered phase formation in FePt thin film. IEEE Trans Magn 54:1–4
11. Goyal R, Lamba S, Annapoorni S (2019) Modelling of strain induced magnetic anisotropy in Au additive FePt thin films. Prog Nat Sci Mater Int 29:517–524
12. Granz SD, Barmak K, Kryder MH (2012) Granular L10 FePt-B and FePt-B-Ag (001) thin films for heat assisted magnetic recording. J Appl Phys 111:07B709
13. Gruber W, Chakravarty S, Baehtz C et al (2011) Strain relaxation and vacancy creation in thin platinum films. Phys Rev Lett 107:265501
14. Hono K, Takahashi YK, Ju G et al (2018) Heat-assisted magnetic recording media materials. MRS Bull 43:93–99
15. Hsiao SN, Yuan FT, Chang HW et al (2009) Effect of initial stress/strain state on order-disorder transformation of FePt thin films. Appl Phys Lett 94:232505
16. Hsiao SN, Chen SK, Liu SH et al (2011) Effect of annealing process on residual strain/stress behaviors in FePt thin films. IEEE Trans Magn 47:3637–3640
17. Hsiao SN, Liu SH, Chen SK et al (2012) Effect of intrinsic tensile stress on (001) orientation in L 10 FePt thin films on glass substrates. J Appl Phys 111:07A702
18. Hsu YN, Jeong S, Laughlin DE et al (2003) The effects of Ag underlayer and Pt intermediate layers on the microstructure and magnetic properties of epitaxial FePt thin films. J Magn Magn Mater 260:282–294
19. Kief MT, Victora RH (2018) Materials for heat-assisted magnetic recording. MRS Bull 43:87–92
20. Kim JS, Koo YM, Shin N (2006) The effect of residual strain on (001) texture evolution in FePt thin film during postannealing. J Appl Phys 100:093909
21. Kim JS, Koo YM, Lee BJ et al (2006) The origin of (001) texture evolution in FePt thin films on amorphous substrates. J Appl Phys 99:053906
22. Kim JS, Koo YM (2008) Thickness dependence of (001) texture evolution in FePt thin films on an amorphous substrate. Thin Solid Films 516:1147–1154
23. Kim CS, Sapan JJ, Moyerman S et al (2010) Thickness and temperature effects on magnetic properties of $L1_0$- ordered FePt films. IEEE Trans Magn 46:2282–2285
24. Leistner K, Thomas J, Schlörb H et al (2004) Highly coercive electrodeposited FePt films by postannealing in hydrogen. Appl Phys Lett 85:3498–3500
25. Li X, Liu B, Sun H et al (2008) L10 phase transition in FePt thin films via direct interface reaction. J Phys D Appl Phys 41:235001

26. Liu M, Jin T, Hao L et al (2015) Effects of Ru and Ag cap layers on microstructure and magnetic properties of FePt ultrathin films. Nanoscale Res Lett 10:1–8

27. Lyubina J, Rellinghaus B, Gutfleisch O et al (2011) Structure and magnetic properties of L10-ordered Fe–Pt alloys and nanoparticles. In: Handbook of magnetic materials, vol 19. Elsevier, pp 291–407

28. Maret M, Brombacher C, Matthes P et al (2012) Anomalous x-ray diffraction measurements of long-range order in (001)-textured L 1 0 FePtCu thin films. Phys Rev B 86:024204

29. PDF Card № 03-065-9121, Menshikov A, Tarnóczi T, Kren E (1975) Magnetic structure of ordered FePt and Fe3 Pt alloys. Phys Stat Sol (a) 28(1):K85–K87

30. Minh PTL, Thuy NP, Chan NTN (2004) Thickness dependence of the phase transformation in FePt alloy thin films. J Magn Magn Mater 277:187–191

31. Nakaya M, Kanehara M, Yamauchi M et al (2007) Hydrogen-induced crystal structural transformation of FePt nanoparticles at low temperature. J Phys Chem C 111:7231–7234

32. Pavlova OP, Verbitska TI, Vladymyrskyi IA et al (2013) Structural and magnetic properties of annealed FePt/ag/FePt thin films. Appl Surf Sci 266:100–104

33. Platt CL, Wierman KW, Svedberg EB et al (2002) L–1 0 ordering and microstructure of FePt thin films with Cu, Ag, and Au additive. J Appl Phys 92:6104–6109

34. Rasmussen P, Rui X, Shield JE (2005) Texture formation in FePt thin films via thermal stress management. Appl Phys Lett 86:191915

35. Takahashi YK, Hono K (2005) On low-temperature ordering of FePt films. Scr Mater 53:403–409

36. Vladymyrskyi IA, Karpets MV, Katona GL et al (2014) Influence of the substrate choice on the L10 phase formation of post-annealed Pt/Fe and Pt/Ag/Fe thin films. J Appl Phys 116:044310

37. Wang F, Xu X, Liang Y et al (2009) FeAu/FePt exchange-spring media fabricated by magnetron sputtering and postannealing. Appl Phys Lett 95:022516

38. Wang B, Barmak K (2011) Re-evaluation of the impact of ternary additions of Ni and cu on the A1 to L10 transformation in FePt films. J Appl Phys 109:123916

39. Wang LW, Shih WC, Wu YC et al (2012) Promotion of [001]-oriented L10-FePt by rapid thermal annealing with light absorption layer. Appl Phys Lett 101:252403

40. Weller D, Parker G, Mosendz O, Lyberatos A, Mitin D, Safonova NY, Albrecht M (2016) Review article: FePt heat assisted magnetic recording media. J Vac Sci Technol, B: Nanotechnol Microelectron: Mater, Process, Meas, Phenom 34(6):060801

41. Wierman KW, Platt CL, Howard JK et al (2003) Evolution of stress with L1 0 ordering in FePt and FeCuPt thin films. J Appl Phys 93:7160–7162

42. Wu AQ, Kubota Y, Klemmer T et al (2013) HAMR areal density demonstration of 1+ Tbpsi on spinstand. IEEE Trans Magn 49:779–782

43. Yamauchi M, Okubo K, Tsukuda T et al (2014) Hydrogen-induced structural transformation of AuCu nanoalloys probed by synchrotron X-ray diffraction techniques. Nanoscale 6:4067–4071

44. Yang WS, Sun TH, Chen SC et al (2019) Comparison of microstructures and magnetic properties in FePt alloy films deposited by direct current magnetron sputtering and high power impulse magnetron sputtering. J Alloys Compd 803:341–347

45. You CY, Takahashi YK, Hono K (2006) Particulate structure of FePt thin films enhanced by Au and Ag alloying. J Appl Phys 100:056105

46. Yu YS, Li HB, Li WL et al (2008) Low-temperature ordering of L10 FePt phase in FePt thin film with AgCu underlayer. J Magn Magn Mater 320:L125–L128

47. Yu YS, Li HB, Li WL et al (2010) Structure and magnetic properties of magnetron-sputtered [(Fe/Pt/Fe)/Au] n multilayer films. J Magn Magn Mater 322:1770–1774

48. Yuan FT, Chen SK, Liao WM et al (2006) Very high coercivities of top-layer diffusion Au/FePt thin films. J Magn Magn Mater 304:e109–e111

49. Yuan FT, Liu SH, Liao WM et al (2012) Ordering transformation of FePt thin films by initial stress/strain control. IEEE Trans Magn 48:1139–1142

50. Zhang WY, Shima H, Takano F et al (2009) Enhancement in ordering of Fe 50 Pt 50 film caused by Cr and Cu additives. J Appl Phys 106:033907

51. Zhang L, Takahashi YK, Perumal A et al (2010) L10-ordered high coercivity (FePt) Ag–C granular thin films for perpendicular recording. J Magn Magn Mater 322:2658–2664
52. Zhang AM, Chen ZX, Zou WQ et al (2012) Effects of substrate on structure and the magnetic properties of (001)-textured FePt films grown at low temperature. J Appl Phys 111:07A704

Chapter 6
New Applications of Spin-Crossover Complexes: Microwave Absorption, Chirooptical Switching and Enantioselective Detection

Olesia I. Kucheriv, Viktor V. Oliynyk, Volodymyr V. Zagorodnii, Vilen L. Launets, Igor O. Fritsky, and Il'ya A. Gural'skiy

Abstract Spin transition complexes form a group of switchable materials that can change their magnetic, optical, electric, mechanical and other physical properties under the influence of different external stimuli. New application areas of such materials keep constantly emerging. Here we describe several different developing applications of spin-crossover materials: (a) the change of microwave absorption as a result of spin transition; (b) switchable chirooptical properties of spin-crossover materials exhibiting molecular asymmetry; (c) enantioselective detection via effect of guest chiral molecules on the transition characteristics of chiral spin-crossover frameworks.

Keywords Spin crossover · Switchable materials · Microwave absorption · Chiral coordination compounds · Enantioselective detection

6.1 Introduction

The study of spin-crossover (SCO) phenomenon forms an important sub-area of a much larger research field of coordination chemistry. SCO is a phenomenon of electronic reorganization between low-spin (LS) and high-spin (HS) states, which can be observed in $3d^4 - 3d^7$ metal complexes under the influence of different external stimuli such as change of temperature [1] or pressure [2, 3], magnetic field

O. I. Kucheriv · I. O. Fritsky · I. A. Gural'skiy (✉)
Faculty of Chemistry, Taras Shevchenko National University of Kyiv, Kyiv, Ukraine
e-mail: illia.guralskyi@univ.kiev.ua

V. V. Oliynyk · V. V. Zagorodnii · V. L. Launets
Faculty of Radio Physics, Electronics and Computer Systems, Taras Shevchenko National University of Kyiv, Kyiv, Ukraine

© Springer Nature B.V. 2020   119
A. Kaidatzis et al. (eds.), *Modern Magnetic and Spintronic Materials*, NATO
Science for Peace and Security Series B: Physics and Biophysics,
https://doi.org/10.1007/978-94-024-2034-0_6

[4], light irradiation [5], or effect of guest molecules [6, 7]. The most important fact is that the spin transition is accompanied by a drastic change of the whole set of physical properties of the molecule such as magnetic [8], electric [9], optical [10], mechanical [11] etc. It has attracted considerable interest of scientists addressing issues of chemistry, physics and biology during the several past decades due to the unique effects of bistability, a variety of switchable characteristics, relevantly low cost and easy ways of modification.

The effect of thermally-induced SCO was firstly observed almost 90 years ago by Cambi et al. [8] in his work on the unusual magnetic behavior of dithiocarbamato complexes of Fe(III). It was shown that in a series of similar complexes with different substituents some showed magnetic susceptibility corresponding to five unpaired electrons (HS state), some showed magnetic susceptibility corresponding to one unpaired electron (LS state), while the third group exhibited HS state at room temperature and switched to the LS state at low temperatures.

Later, the first Fe^{II} SCO compound was discovered. It was [Fe(phen)$_2$(SCN)$_2$] (phen = 1,10-phenatroline), exhibiting an abrupt thermally induced spin transition [12, 13]. Since then SCO was found in many other metal complexes, such as Co^{II}, Co^{III}, Cr^{II}, Mn^{II}, Mn^{III} and Ni^{III} [14–16].

Here we focus on the temperature induced spin transition phenomenon in Fe^{II} complexes, including a brief overview of selected classes of SCO compounds. In addition, we go to a deeper analysis of currently developing directions of SCO research. New trends in chiral coordination compounds and their applications are discussed in detail. Different ways to induce chirality of a SCO coordination framework and some examples of chiral SCO complexes known for today are shown. Furthermore, an emerging research area of microwave switching by means of SCO materials is introduced.

6.2   Spin Transition in Coordination Materials

The spin transition phenomenon is explained by the ligand field theory, which is the sequent of the crystal field theory that was formulated in 1929 by Bethe [17] almost at the time with the discovery of SCO. In octahedral environment five nd orbitals split into two sublevels: t_{2g} which is formed by d_{xy}, d_{yz}, d_{zx} orbitals and e_g which is formed by d_{z^2} and $d_{x^2-y^2}$ orbitals. The t_{2g} orbitals are practically non-bonding and consequently have lower energy than the anti-bonding e_g orbitals. This energy difference between the two sets is described by the crystal field splitting parameter, Δ_{oct}, which is determined by the nature of a ligand, its way of coordination, and the metal itself.

Electron-electron repulsion (which can be described as spin paring energy) should be considered except for ligand field strength in case of a system with more than one d-electron. For an ion with d^6 electron configuration, for example Fe^{2+}, two extreme cases of placing electrons on t_{2g} and e_g orbitals are possible. When the

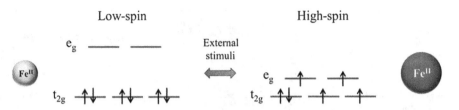

Fig. 6.1 Two possible electronic configurations of a $3d^6$ metal ion

ligand field is weak comparing to the spin pairing energy, the electrons occupy the five d orbitals ensuring the maximum spin multiplicity, that gives rise to the HS ground state $^5T_{2g}$ ($t_{2g}^4 e_g^2$) (Fig. 6.1). On the contrary, when the ligand field is strong, the electrons pair on the bonding t_{2g} orbitals that stabilizes the LS ground state $^1A_{1g}$ (t_{2g}^6). The $^5T_{2g}$ state remains the ground state until the critical ligand field is reached. This happens when Δ_{oct} is equal to the spin paring energy. Above this value the $^1A_{1g}$ state becomes the ground state, and the HS form is stabilized. During the spin transition metal-ligand bond length changes due to the transfer of two electrons, for example from 1.95 to 2.00 Å in the LS form to ~2.2 Å in the HS form in Fe^{II} compounds. The ligand field strength changes as well, because it depends on the metal-ligand distance as $1/r^n$, with n = 5–6. Deeper consideration of spin transition process in terms of ligands filed theory is given in Ref. [10].

The majority of SCO complexes known for today are formed by metals with d^6 configuration, these are mostly Fe^{II} complexes. These complexes undergo the transition between fully diamagnetic LS state with spin multiplicity S = 0 and paramagnetic HS state with S = 2 (Fig. 6.1). The spin transition in the Fe^{II} complexes is followed by the distinct color change, e.g. colorless ↔ pink [18], orange ↔ red [19], red ↔ blue [20]. The HS state is stabilized at higher temperatures and the transition to the LS state occurs upon cooling to lower temperatures.

The principal technique for characterization of spin transition in coordination materials is a measurement of magnetic susceptibility as a function of temperature $\chi(T)$. In case of Fe^{II} compounds, the transition from diamagnetic LS state to paramagnetic HS state results in a drastic change of χ. Thus, such important information as temperature, abruptness, completeness and width of hysteresis loop can be derived from temperature dependent magnetic curves of SCO materials.

The example of a magnetic curve of an abrupt hysteretic spin transition in a well-known [Fe(Htrz)(trz)]BF$_4$ (Htrz = 1,2,4-triazole, trz = triazolato anion) [21] complex and a SCO induced thermochromic effect is shown in Fig. 6.2.

There are several typical classes of SCO compounds that are formed by different N-donor ligands. One of the biggest classes is represented by complexes of Fe^{II} with 4-R-1,2,4-triazoles and different anions of general formula [Fe(4-R-1,2,4-triazole)$_3$] A$_x$ [18, 22]. These complexes attract much attention due to their low cost, simple ways of synthesis, and easy approaches of modification by the change of a substituent in a triazole ring or anion. Additionally, a lot of complexes of this family exhibit spin transition at practically attractive high temperatures or even at room temperature [23]. These complexes have a one-dimensional polymeric structure formed by

Fig. 6.2 Example of a
magnetic curve of Fe^{II}
complex displaying an
abrupt hysteretic spin
transition; (*inset*)
thermochromic effect
associated with spin
transition

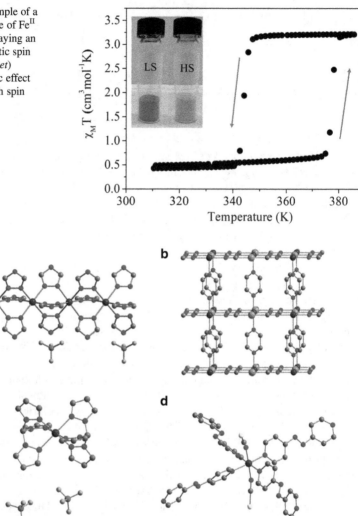

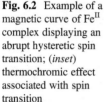

Fig. 6.3 Crystal structures of four representatives that belong to selected classes of SCO compounds: (**a**) 4-R-1,2,4-triazole based complex, (**b**) complex which is a Hofmann clathrate analogue, (**c**) tris-pyrazolylborate based complex, (**d**) complex of [Fe(NCX)2(N-donor ligand)n] type

Fe-triazolic chains. Counter anions are located in the cavities between the polymeric chains providing the linkage to supramolecular 3D structure (Fig. 6.3a) [24, 25]. For details regarding structure and magnetic behaviors of 4-R-1,2,4-triazole based complexes some reviews are recommended [18, 22, 26].

Another popular class of SCO complexes is formed by Hofmann clathrate analogues. These complexes are 2D and 3D coordination polymers that are formed by heterocyanometallic layers and N-donor heterocyclic ligands of general formula

[Fe(L)$_x${M(CN)$_y$}$_z$]. The first clathrate of this type was obtained in 1897 by Hofmann and Kuspert. It was a [Ni(NH$_3$)$_2${Ni(CN)$_4$}] polymeric framework formed by infinite [Ni{Ni(CN)$_4$}] layers supported by NH$_3$ ligands which could incorporate small aromatic guest molecules [27]. In 1996 Kitazawa et al. received the first SCO Hofmann clathrate analogue with general formula [Fe(py)$_2${Ni(CN)$_4$}] [28].

Numerous new SCO complexes of Hofmann clathrate type were obtained since then. The modification of chemical composition is usually performed by introduction of different cyanometallic anions, such as square-planar [M(CN)$_4$]$^{2-}$ (M = Ni, Pt, Pd) [19, 28, 29], linear [M(CN)$_2$]$^-$ (M = Cu, Ag, Au) [30, 31] or dodecahedral [M(CN)$_8$]$^{4-}$ (M = Nb, W) [32]. Furthermore, numerous N-donor organic heterocyclic ligands were used to obtain new Hofmann clathrate analogues, for example different substituted and non-substituted [19, 33] azines. Additionally, the inclusion of bidentate aromatic ligands such as pyrazine allowed the formation of a 3D supramolecular frameworks (Fig. 6.3b). The most prominent examples of Hofmann clathrate analogues known for today include complexes with multistep [34–36] or room temperature transitions [19, 37]. For more details on the topic of Hofmann clathrate analogues some reviews are recommended [6, 7, 38].

Tris(pyrazol-1-yl)methane and tris(pyrazol-1-yl)borate complexes form another family of SCO materials [39]. They are neutral and cationic mononuclear complexes of general formula [Fe{HB(pz)$_3$}$_2$] or [Fe{HC(pz)$_3$}$_2$]A$_2$ (pz = pyrazine). These ligands are usually coordinated with Fe^{II} in a tridentate mode as shown in Fig. 6.3c. The modification of these complexes can be performed via introduction of various substituents or different anions in case of tris(pyrazol-1-yl) methane, also by introduction of triazole ring instead of pyrazole. An important feature of some of the complexes from this class is their volatility. This property allowed preparation of ultrathin SCO thin films of high quality [40].

In addition, SCO complexes of general formula [Fe(NCX)$_2$(N-donor aromatic ligand)$_n$] (X = S, Se) [41] are quite widespread [42]; their typical crystal structure is shown in Fig. 6.3d. Complexes which belong to this family tend to form 2D porous scaffolds with guest accessible voids that allows their use in molecular sensing [43].

Thanks to the variety of SCO complexes with different transition characteristics and structural features, numerous applications of these materials have been offered nowadays. For example, SCO materials were shown to be efficient molecular sensors [6, 7]. This effect is associated with very high sensitivity of SCO materials to the inclusion of some guest molecules, upon which a drastic change of magnetic properties can occur. For example, a shift of transition temperature, change of abruptness or even appearance or disappearance of spin transition have been observed as a result of inclusion of some guest molecules [44].

Recent achievements in the study of molecular SCO offer numerous other potential applications of these materials (Fig. 6.4): molecular actuation [11, 45], microthermometry [46], anti-counterfeit elements [47], memory devices [48] etc.

Various reviews showing new aspects in the design of new Fe^{II} based SCO materials and their properties are constantly being published. For example, the questions of SCO in complexes with different ligand systems [49–53], guest effect

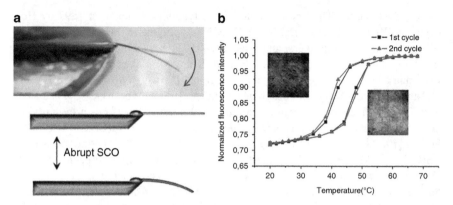

Fig. 6.4 Some prominent examples of SCO materials application: (**a**) molecular actuators; (**b**) anti-counterfeit thermofluorescent cellulose nanocomposites. (Images are adapted from Refs. [11, 47] with permission. Copyrights 2013, Springer Nature; 2014, Elsevier B.V)

[6, 7, 54], SCO nanostructures [55], new potential applications [56, 57], and many other [58–63] have been recently reviewed.

Here we describe some developing directions of research in the field of SCO materials: microwave switching properties, chiral materials, and enantioselective detection.

6.3   Chirality in SCO Materials

The effect of molecular asymmetry, which is in general case more common for organic materials, becomes more and more demanded in the field of inorganic and coordination chemistry. Introduction of chiral centers allows to obtain new coordination compounds with specific physical properties such as circular dichroism (CD), chiroptical switching [64], nonlinear optical properties, circularly polarized luminescence, etc. Chiral coordination compounds were employed for the development of enantioselective sensors, chiral magnets, chiral switching [65] etc.

Molecular switches are materials that can exist in two different stable states and reversibly transit between them by application of external stimuli. Different ways to switch chiral coordination materials were shown for today. For example, chiral propeller-shaped complexes were shown to exhibit reversible switching between two states as a result of redox reaction [64]. In addition, switching between two states with different optical rotations was performed by reversible photocyclization in copper complexes with 1,2-dithienylethenes derivatized with chiral oxazoline moieties [66]. In this context, SCO coordination materials that contain asymmetric centers represent huge interest for the development of chiral molecular switches. Around 30 chiral SCO compounds are known for today [67–94], they were obtained by different synthetic approaches using diverse ligand systems and thus exhibit spin transitions with different temperatures, completeness, abruptness, and hysteresis.

6.3.1 Types of Chiral SCO Complexes

There are two main approaches to induce chirality of a SCO framework: by obtaining chiral SCO crystals from non-chiral precursors and subsequent separation or using chiral precursors with predefined chirality. Both of them have been successfully used for the production of chiral SCO frameworks.

6.3.1.1 SCO Complexes which Crystallize in Chiral Groups

One of the first examples of chiral SCO compounds was offered by Matsumoto et al. in 2003 [68]. A mixed valence $[Fe^{II}H_3L][Fe^{III}L](NO_3)_2$ (H_3L = tris{[2-{(imidazole-4-yl)methylidene}amino]ethyl}amine)) complex was obtained by a one-pot self-assembly of $[Fe^{II}H_3L]^{2+}$ and in situ generated $[Fe^{III}L]^{3+}$ moieties. The chirality of the offered material is determined by coordination of achiral tripoid ligand in a clockwise or anticlockwise manner. Temperature induced spin crossover in Fe^{II} and Fe^{III} centres of this complex allowed observation of three different electronic states: LS Fe^{II} – LS Fe^{III}, HS Fe^{II} – LS Fe^{III} and HS Fe^{II} – HS Fe^{III}.

Later similar SCO complexes that exhibit spontaneous resolution upon crystallization were obtained with this ligand and tetrafluoroborate anion. The controlled deprotonation of a ligand allowed to obtain three different complexes: $[Fe^{II}H_3L]$ $(BF_4)_2$, $[Fe^{II}H_3L][Fe^{II}L]BF_4$ and $[Fe^{II}H_3L][Fe^{III}L](BF_4)_2$ that offer a rich variety of SCO behaviors [73]. Later on it was shown that chiral SCO formally hemideprotonated Fe^{3+} complex of this series $[Fe^{III}(H_3L)][Fe^{III}(L)]^{3+}$ can selectively co-crystalize with $[Cr^{III}(ox)_3]^{3-}$ complex [91]. Chiral voids constructed by the host Δ-Fe^{III} components enantioselectively discriminate against $[Cr^{III}(ox)_3]^{3-}$ complex creating sheets which contain Δ-$[Fe^{III}(H_3L)]^{3+}$, Δ-$[Fe^{III}(L)]$, and Λ-$[Cr^{III}(ox)_3]^{3-}$ moieties.

Similar tridentate ligand ((2-methylimidazol-4-yl)methylidene)histamine (abbreviated as H_2L^{2-Me}) gave rise to the complex $[Fe(H_2L^{2-Me})_2]Cl_2 \cdot 2$-$PrOH \cdot 0.5H_2O$ which crystalizes in enantiopure forms and exhibits an abrupt spin transition at 180 K (in desolvated form) [92, 93].

Tong et al. demonstrated a chiral metal-organic framework exhibiting SCO $[Fe^{II}(mptpy)_2] \cdot EtOH \cdot 0.2DMF$ (Hmptpy = 3-methyl-2-(5-(4-(pyridin-4-yl)phenyl)-4H-1,2,4-triazol-3-yl)-pyridine) which was solvothermally synthesized through spontaneous resolution. This complex exhibits a two-step SCO with a plateau at high temperature (T_{c1} = 200 K, T_{c2} = 357 K) [90].

A bimetallic cyano-bridged Fe-Nb chiral SCO metal-organic framework was offered by Ohkoshi et al. [74] Complex $Fe_2[Nb(CN)_8](4$-bromopyridine$)_8 \cdot 2H_2O$ crystallizes in $I4_122$ space group creating a helical structure along the fourfold

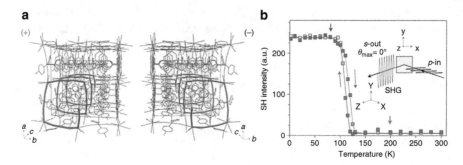

Fig. 6.5 (a) Crystal structure of $Fe_2[Nb(CN)_8](4\text{-bromopyridine})_8\cdot 2H_2O$ showing its helical nature; (b) temperature dependence of the second harmonic intensity of a single crystal of $Fe_2[Nb(CN)_8](4\text{-bromopyridine})_8\cdot 2H_2O$ which drastically changes during the spin transition. (Images reprinted with permission from Ref. [74]. Copyright 2013, Springer Nature)

screw axis (Fig. 6.5). This compound was shown to be a unique material which combines three phenomena: spin-crossover induced second harmonic generation, light-reversible spin-crossover long-range magnetic ordering, and photoswitching of magnetization-induced second-harmonic generation.

Other ligand systems suitable for creation of chiral SCO frameworks were offered by Real et al. This team described the first SCO complex based on singular C_3-symmetric tris-bidentate bridging ligand 1,4,5,8,9,12-hexaazatriphenylene (HAT) – $[Fe(HAT)(NCS)_2]_\infty\cdot nMeOH$. This homochiral complex exhibits a complete two-step spin transition. Other chiral SCO materials offered by this team were $[Fe^{II}(bqen)(NCX)_2]$ (X = S, Se, bqen = N,N′-bis(8-quinolyl)ethane-1,2-diamine) complexes [87]. They were obtained in a form of two polymorphs one of which was a racemic Λ-Δ mixture of complexes, while the other polymorph contained only one enantiomer. All four complexes undergo spin transition that allowed making correlations between SCO properties of enantiopure and racemic polymorphic forms.

Thus, spontaneous crystallization allows to obtain chiral SCO materials based on different ligand systems. These complexes exhibit various SCO behaviours and other fascinating physical properties. At the same time, this method does not allow to produce higher quantities of enantiopure complexes or to study non monocrystalline forms of chiral materials exhibiting spin transition.

6.3.1.2 SCO Complexes with Chiral Ligands

One of the ways to introduce asymmetric centers with predefined chirality into a SCO system is the use of chiral ligands. For example, Matsumoto et al. obtained a SCO complex with S and R forms of chiral bidentate chelate ligand 2-methylimidazol-4-yl-methylideneamino-1-methylphenyl (HL). The R-form of the ligand induces the creation of a fac-Λ-isomer of $[Fe^{II}(HL)_3](ClO_4)_2$ complex, and the S-form of the ligand forms a fac-Δ-isomer, because of the steric requirements. The fac-Λ-isomer was shown to exhibit an abrupt spin transition with $T_c = 195$ K [69].

A Schiff base ligand was used by Gu et al. to obtain a pair of mononuclear iron (II) enantiomeric complexes: fac-Λ-[Fe(R-L)$_3$](BF$_4$)$_2$·MeCN and fac-Δ-[Fe(S-L)$_3$] (BF$_4$)$_2$·MeCN where L = 1-phenyl-N-(1-methyl-imidazol-2-ylmethylene) ethanamine. Magnetic measurements showed that the complexes display the same spin transition properties with $T_{up} = 222$ K and $T_{down} = 219$ K [83].

This team has made a big contribution into the field of chiral SCO materials by developing a whole family of complexes by chemical modification of basic components: 1-imidazole and 4-phenylethylamine substitution [81, 86, 88] or by introduction of 1-(2-naphthyl)-ethylamine instead of phenylethylamine [79].

In addition, fac-Λ-[FeL$_3$](ClO$_4$)$_2$ was used for optical recognition of alkyl nitrile. It was shown that when the R isomer of the complex is dissolved in racemic lactonitrile (LN) or methylglutaronitrile (MGN), chiral [FeL$_3$]$^{2+}$ moieties can recognize and selectively crystalize with one of the isomers forming fac-Λ-[FeL$_3$] (ClO$_4$)$_2$·1/3(R)-LN or fac-Λ-[FeL$_3$](ClO$_4$)$_2$·1/3(S)-MGN solvates. Both solvates exhibit incomplete spin transition at 363 K [80].

A nice example of multifunctional chiral SCO material is a coordination polymer [Co^{III}((R)-pabn)][Fe^{II}(tp)(CN)$_3$](BF$_4$)·H$_2$O (R-pabn = (R)-N(2),N(2')-bis(pyridin-2-ylmethyl)-1,1'-binaphtyl-2,2'-diamine, tp = hydrotris(pyrazol-1-yl)borate). This bimetallic [CoFe] cyanide-bridged chiral system exhibits an abrupt thermal, light-induced and electron-transfer coupled spin transition as well as thermal conductivity switching between semiconducting (HT) and insulating (LT) phases and light induced single chain magnet properties (Fig. 6.6a,b) [95].

In 2015 Pilkington et al. offered a macrocyclic complex employing a chiral ligand obtained by Schiff-base condensation of R,R-(+)-diamine and 2,6-diacetyl pyridine. This complex crystalizes with two independent molecules one of which has [FeO$_2$N$_5$] and the other [FeON$_5$] coordination environment. Spin transition in this complex was shown to be gradual and incomplete [82]. A multicomponent self-assembly process of simple flexible di(imidazole aldehyde), chiral phenylethylamine and Fe^{II} ions leads to the creation of chiral SCO tetrahedral cages [78]. These complexes exhibit spin-crossover close to room temperature.

In 2017 Zuo et al. reported on the fabrication of four enantiopure Fe^{II} complexes [Fe(R-L^1)(bpz)$_2$], [Fe(S-L^1)(bpz)$_2$] (L = 2,3-bipyridne or 1,10-phenentroline, bpz = bis(1-pyrazolyl)borohydride). These complexes exhibit an abrupt one-step spin transition which is accompanied by a major change in the dielectric constant [76]. Complexes of [Fe(L)$_2$(NCS)$_2$] type (with L = α-methyl-N-(2-pyridinylmethylene)-cyclohexanemethanamine) were obtained in pure enantiomeric and racemic forms. They were shown to display a gradual incomplete spin transition [75]. Another family of chiral SCO complexes is based on 2,6- bis (oxazolinyl)pyridine (PyBox) derivatives. Several complexes of this type with different substituents were shown to display variable SCO behaviours including abrupt and multistep transitions [71, 77].

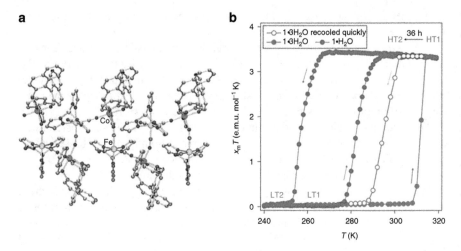

Fig. 6.6 (**a**) Crystal structure of polymer [Co^{III}((R)-pabn)][Fe^{II}(tp)(CN)$_3$](BF$_4$)·H$_2$O (R-pabn = (R)-N(2),N(2′)-bis(pyridin-2-ylmethyl)-1,1′-binaphtyl-2,2′-diamine, tp = hydrotris(pyrazol- 1-yl) borate). Counteranions, hydrogen atoms and solvent molecules are omitted for clarity; (**b**) temperature dependence of $\chi_M T$ for this complex showing an abrupt temperature induced spin transition. (Images are reprinted with permission from ref. [95]. Copyright 2012, Springer Nature)

6.3.1.3 SCO Complexes with Chiral Anions

Another approach which was effectively used to ensure the chirality of SCO frameworks is introduction of chiral anions. One of the most recent examples of SCO materials obtained with chiral anions was offered by Morgan et al. in 2019. The introduction of a R,R or S,S isomer of chiral anion bis[1,1′-binaphthyl-2,2′-diolato] boron (X) allowed to resolve enantiomeric Mn^{3+} cations and to obtain enantiopure Δ-[MnL]{(S,S)–X}·solvent or Λ-[MnL]{(R,R)–X}·solvent complexes with a Schiff base hexadentate ligand [96]. Both enantiomers exhibit a gradual incomplete thermal spin crossover.

Chiral Δ-As$_2$(tartrate)$_2$ anion was used to synthesize optically active [Fe(phen)$_3$] [Δ-As$_2$(tartrate)$_2$] complex which was shown to undergo ultrafast LS to HS photoswitching upon excitation of a thin film of this complex by a femtosecond laser pulse [70, 72].

Different approach was used by Kinizuka et al. who obtained a Fe^{II} – 4-amino-1,2,4-triazole complex where chiral L-glutamate-derived lipid was introduced as a lipophilic counteranion.

Our team offered a new homochiral SCO complex containing a chiral anion [84]. We showed a route towards production of a 1D triazolic complex [Fe (NH$_2$trz)$_3$](L-CSA)$_2$ (NH$_2$trz = 4-amino-1,2,4-triazole, L-CSA = L-camphorsulphonate) in forms of nanoparticles and gel (Fig. 6.7). The nanoparticles were obtained as a stable colloid by reaction of the corresponding salt and ligand in acetonitrile using a surfactant free approach. The spin transition in chiral SCO

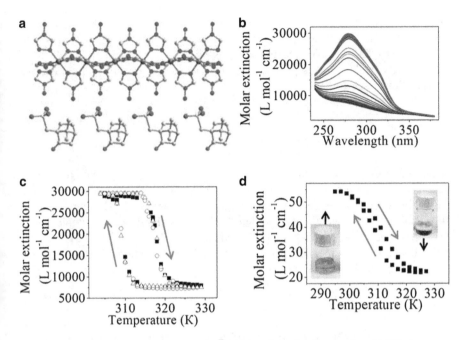

Fig. 6.7 (a) A schematic representation of Fe^{II} 4-NH$_2$–1,2,4,-triazolic complex; (b) UV-vis spectra of colloidal solution of [Fe(NH$_2$trz)$_3$](L-CSA)$_2$ nanoparticles, measured in the 304–329 K temperature range; (c) molar extinction vs. temperature curve for nanoparticles obtained by monitoring MLCT band; (d) molar extinction vs. temperature curve for gel obtained by monitoring d-d transfer band

nanoparticles was monitored by UV-vis spectroscopy by following the temperature induced decrease of metal-to-ligand charge transfer (MLTC) band in the UV region. The band has a high extinction coefficient ($\varepsilon = 30,000$ L mol^{-1} cm^{-1}) when the complex is in the LS form, which decreases during the conversion to the HS state. The measurements revealed that an abrupt spin transition in the nanoparticles takes place at 318 K upon heating, in the cooling mode the nanoparticles switch back to the LS form at 309 K demonstrating the cooperativity of the process ($\Delta T = 9$ K).

The magnetic measurements, conducted after the removal of the solvent, showed a typical solvent induced change of transition temperatures. In this case, spin transition in dry complex takes place at 316 K in heating mode and at 300 K in cooling mode. The values of $\chi_M T$ below and above spin transition are typical for the LS and HS forms, respectively, and indicate that the spin transition is complete.

In addition, our team used this complex for the preparation of a chiral SCO gel. The gel was also obtained by direct reaction of corresponding salt and ligand, but in this case chloroform was used instead of acetonitrile. The formation of this gel proves that despite the general opinion it is unnecessary to have a ligand/anion with long lipophilic chains to obtain a gel. Formation of a gel is indicated by increase of viscosity of a system which is caused by formation of a complex-solvent supramolecular framework. Spin transition of the gel was monitored by the change of

$^1A_1 \rightarrow {}^1T_1$ d-d transition band which can be detected in concentrated solutions. In case of a gel, SCO takes place at similar temperatures (T_{up} = 313 K and T_{down} = 310 K), however the transition is slightly more gradual. The loss of cooperativity in this case is provoked by the change of size and morphology of coordination framework.

6.3.1.4 SCO Complexes with Chiral Cations?

At the same time, to the best of our knowledge, there are no anionic chiral SCO complexes known for today in which the non-SCO cation would provide the chiral center. In general, there are only several examples of anionic SCO complexes, but still they can be potentially modified to become chiral. For example, SCO $(Me_2NH_2)_6[Fe_3(\mu\text{-}L)_6(H_2O)_6]$ [97] (where L = 4-(1,2,4-triazol-4-yl) ethanedisulfonate) complex is a perfect candidate for introduction of a chiral ammine containing cation instead of $Me_2NH_2^+$.

6.3.2 SCO Complexes with Switchable Circular Dichroism

Circular dichroism is a powerful tool for monitoring optically active materials. It detects the differential absorption of left and right circularly polarized light. This approach is frequently used to study important biological molecules, as their dextrorotary and levorotary components display differential CD signals. Moreover, even secondary structure of proteins and DNA can be determined by CD spectroscopy [98]. In addition, CD spectroscopy appears to be preferential for the study of chiro-optical properties of coordination compounds.

CD spectra can be detected only when the system contains a chromophore with a chiral center. In case of SCO compounds, CD spectra measured at variable temperatures in the temperature range of SCO attract considerable attention, as they allow to monitor the switching of chiro-optical properties induced by the spin transition. This technique was used [67, 84] for triazolic SCO complexes with high temperature SCO.

In case of $[Fe(NH_2trz)_3](L\text{-}CSA)_2$ complex, at low temperatures, when the complex is in the LS form (308 K) five CD bands are observed: three negative (344, 293 and 278 nm) and two positive (317 and 247 nm). These bands originate from iron to triazole MLCT and n $\rightarrow$ π^* transition in camphorsuphonate anion (Fig. 6.8a).

Upon heating, when the transition to the HS state occurs, these bands disappear and two new emerge: positive at 301 nm and negative at 255 nm. Additionally, spin transition in the complex can be followed by monitoring the temperature induced changes of CD bands (Fig. 6.8c).

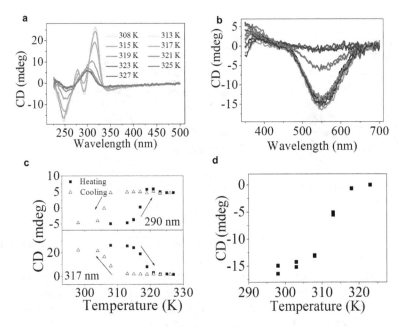

Fig. 6.8 CD spectra of [Fe(NH$_2$trz)$_3$](L-CSA)$_2$ nanoparticles (**a**) and gel (**b**) measured in the 293–323 K temperature range; CD vs. temperature dependence of nanoparticles (**c**) and gel (**d**)

Temperature dependent CD spectra of the gel are shown in Fig. 6.8b,d. Similar to nanoparticles, a band, which is characteristic for the LS form (negative at 540 nm), disappears when the complex switches to the HS state.

These drastic changes of the chiroptical properties of this material during the spin transition are associated with two interconnected factors: electronic reorganization of the molecule and consequent structural changes. Thus, while during the spin transition major transformations concern Fe^{II} centres and its coordination environment and not the chiral chromophore itself (in this case an anion), these transformations can indirectly influence chiro-optical response of the framework resulting in a drastic change of circular dichroism.

6.4 Enantioselective Detection by SCO Materials

SCO materials are known to be very sensitive to inclusion of different guest molecules. For example, tunability of SCO behavior has been achieved multiple times for different Hofmann clathrate analogues by absorption of benzene, antracene, tiourea, and many other guest molecules [6, 7]. A the same time, Fe^{II}-1,2,4-triazolic complexes tend to form solvates with different hydroxylic solvents and change their SCO characteristics upon contact with water or alcohols. Such kind

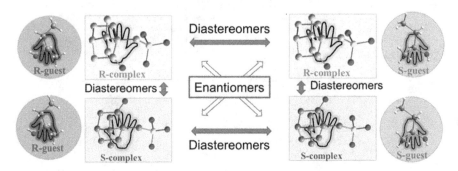

Fig. 6.9 Schematic representation of diastereomeric and enantiomeric [chiral SCO complex] – [chiral guest] pairs

of sensitivity makes chiral SCO complexes being very perspective candidates for enentioselective detection of various chiral guests.

As shown in multiple publications, two enantiomeric forms of the same SCO complex exhibit identical spin transition. This observation well corroborates to general rule concerning all enantiomeric molecules that must have identical chemical and physical properties in an achiral environment. At the same time, introduction of guest molecules with different chiralities to the SCO system (= introduction of second chiral center) will create diasteromeric pairs (Fig. 6.9). As known, diastereomeric molecules have different physical properties (e.g. melting point, solubility, etc.), consequently, such diastereomers should have different spin transition characteristics that will allow enantioselective detection of chiral guest molecules by SCO materials.

Our team showed that two enantiomeric complexes [Fe(NH$_2$trz)$_3$](S-CSA)$_2$ and [Fe(NH$_2$trz)$_3$](R-CSA)$_2$ indeed have identical temperature and abruptness of the spin transition. Upon inclusion of guest molecules of 2-buthanol (racemic mixture) the spin transition becomes more abrupt and shifts to slightly higher temperatures with T_{up} = 324 K and T_{down} = 315 K (Fig. 6.10). This effect is associated with the creation of additional hydrogen bonds in the system that makes spin transition more cooperative [99].

In order to demonstrate enantioselective detection, we obtained four solvates of two enantiomeric complexes and two enantiomeric 2-buthanols (as shown in Fig. 6.9). Their magnetic characteristics were monitored using SQUID magnetometry (Fig. 6.10c). For two diastereomeric solvates a difference in spin transition temperature was observed: solvate of R- complex switches from the LS to the HS state at 325 K with R-BuOH and at 323 K with S-BuOH. Upon cooling HS to LS transition occurs at 316 K with R-BuOH and at 314 K with S-BuOH. This 2 K difference is constant in consequent cycles. Such shift of spin transition temperature in the solvates is a consequence of creation of diastereomeric complex-guest pairs which have two different sets of physical properties including SCO characteristics. The nature of this shift is proved by a completely mirror image of SCO curves obtained for solvates with S complex.

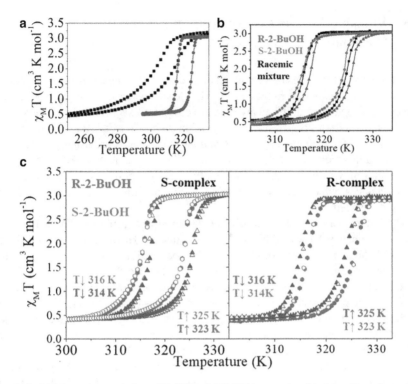

Fig. 6.10 (a) Magnetic curve of pure [Fe(NH₂trz)₃](S-CSA)₂ complex (black) and its solvate with racemic mixture of 2-buthanols (blue); (b) magnetic curves of S-complex solvates with R-, S- and racemic 2-buthanols; (c) magnetic curves of four solvates showing enantioselective shift of spin transition temperatures

The origin of this stereoselectivity was shown to be a differential absorption of R- and S- buthanols: S-complex absorbs 20% more of S-alcohol than R-alcohol. Opposite effect is observed with the R-complex. In this case, a greater absorption of the alcohol induced a bigger shift of the transition temperature. Additionally, the spin states in four solvates were monitored by Mössbauer spectroscopy which proved the stereoselective shift of transition temperatures.

Thus it was shown that the application of SCO material can be widen from usual detection of guest molecules towards stereospecific determination of chiral analytes. Moreover, this method of enantioselective detection can be applied not only via magnetic and Mössbauer experiment, but also through other techniques suitable for the detection of SCO, such as UV-vis, IR, Raman spectroscopies, electric measurements and others.

6.5   Variable Microwave Absorption Properties of SCO Materials

The industry of wireless communication constantly develops simultaneously expanding the frequency ranges used. This leads to the demand for the device components, which could be easily tuned to cover the exact frequency bands.

Different modern systems of communication such as Wi-Fi, GPS, LTE, and so on function in different frequency ranges and all these ranges are usually covered by different radio-frequency components such as filters and antennas. Additionally, the frequencies of wireless standards vary in different parts of the world. Consequently, having devices, which could be reconfigured to tune different frequency bands are highly demanded.

The task of radio switching is solved today by several different approaches, such as radio-frequency MEMS, PIN diodes, transistors, etc. However, some applications need switchable attenuation or reflection of some distributed objects (surfaces etc.) so the use of functional materials is preferable to lumped circuits.

Recently a quite facile way of RF switching has been offered by introduction of phase transition materials such as chalcogenides [100–102] or VO_2 [103, 104]. Taking into account those successful examples of RF switching ability, our team studied the ability of several SCO complexes to change their microwave absorption properties upon temperature induced spin transition [105].

The ability of SCO complexes to change their microwave absorbing properties was studied for two well-known complexes which belong to different classes of SCO compounds: 1,2,4-triazolic complex [Fe(Htrz)$_2$(trz)]BF$_4$ (**1**) [21] and Hofmann clathrate analogue [Fe(pz){Au(CN)$_2$}$_2$] (**2**) [106].

The spin transition curve of **1** obtained by magnetic susceptibility measurement is shown in Fig. 6.11a. It shows that upon heating the transition to the HS state occurs at 377 K. Upon cooling the complex switches back to the LS state at 344 K, revealing 33 K wide hysteresis loop.

Microwave transmission parameters S_{21} were recorded in 26–37.5 GHz frequency range at variable temperatures (Fig. 6.11b). At low temperature the transmission spectra are characterized by the presence of two bands at 28.7 and 30.7 GHz. When the temperature increases, the spin transition in this material causes a change of microwave transmission bands. First of all, the band at 30.7 GHz disappears completely, while the other band changes its intensity and shifts a bit towards lower frequencies.

Temperature dependence of S_{21} measured in heating mode is given in Fig. 6.11c. The transmission coefficient does not vary much at the temperatures below spin transition (295–375 K). But an abrupt change of S_{21} is observed on reaching the transition temperature. Importantly, this parameter can either increase or decrease depending on the operating temperature. For example, S_{21} is at the level of -1.25 dB at 32 GHz and room temperature, but with the transition to the HS state this transmission parameter increases to -0.7 dB. At the same time, the initial value of transmission is -0.9 dB at the frequency of 27 GHz while the spin transition

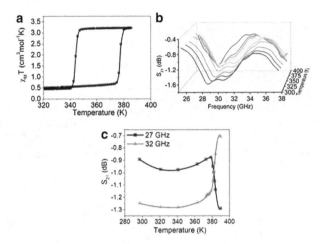

Fig. 6.11 (**a**) Magnetic curve of **1** displaying temperature induced spin transition at $T_{up} = 377$ K and $T_{down} = 344$ K; (**b**) microwave transmission spectra of **1** measured at variable temperatures; (**c**) temperature dependence of S_{21} at selected frequencies

decreases it down to -1.3 dB. This effect arises from the shift of transmission bands towards lower frequencies, which in its turn is caused by an increase of permittivity with the transition to the HS state.

In order to study the nature of SCO induced variation of microwave transmission coefficient, temperature dependent measurements of dielectric parameters were performed at 37 GHz (Fig. 6.12). The experiment was performed using short-circuited waveguide method. In this method the end of a waveguide is blocked with a metallic plate, which creates a standing wave. When the sample is inserted into the waveguide, the measurable parameters of that standing wave change and one can deduce the dielectric characteristics of a material. The real part of permittivity at room temperature equals ca. 1.84, but when the temperature of spin transition is reached, this value increases to 2.02 ($\Delta\varepsilon' = 0.18$). The same trend is observed for imaginary part of permittivity, which increases from 0.042 to 0.050 with the transition from the LS to the HS state. Additionally, there is a possibility to calculate variation of refraction index and absorption of microwave radiation during the spin transition from this data.

The temperature dependent refraction index of **1** is given in Fig. 6.12c. The curve shows the value of refraction index of 1.36 at low temperature. With the temperature increase and transition to the HS state this value increases and reaches 1.42 at 395 K. The same trend is observed for the microwave absorption: an increase of absorption occurs with the LS $\rightarrow$ HS transition. All the parameters revert back to their initial values upon cooling when the complex turns back to the LS state.

Another complex which was chosen to study in microwave frequency range belongs to the family of Hofmann clathrate analogues. This compound is known for its high transition temperatures, stability and reproducibility of spin transition. The spin transition curve of **2** obtained by magnetic measurements is shown in

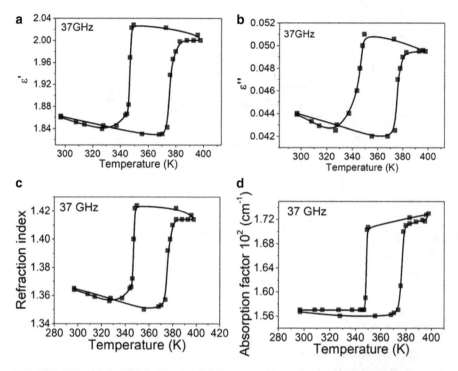

Fig. 6.12 Effect of the SCO in **1** on the dielectric properties: temperature dependence of real (**a**) and imaginary (**b**) parts of permittivity, refraction index (**c**) and microwave absorption (**d**) measured at 37 GHz

Fig. 6.13a. The measurement demonstrates an abrupt spin transition from dia- to paramagnetic state at 367 K and from para- to diamagnetic at 349 K. The color change in the complex induced by spin transition is shown in Fig. 6.13b.

Measurements of the microwave transmission of this complex sample were performed at variable temperatures in 38–55 GHz frequency range in heating mode (Fig. 6.13c). In general, S_{21} varies between -0.8 and -1.3 dB at low temperatures for the LS form of the complex. The transmission increases to $-0.4 - -0.8$ dB with the transition to the HS state. In whole studied frequency range spin transition provokes an increase of microwave transmission. The temperature dependence of S_{21} for **2** at different frequencies is shown in Fig. 6.13d. As well, similar to **1**, refraction index and permittivity of **2** measured at 39 GHz increases during the transition from the LS to the HS state [107].

In general, the permittivity of complexes at different frequency ranges is determined by different mechanisms of polarizability. Thereby, permittivity can either increase or decrease at different frequencies with the transition to the HS state. For example, an increase of dielectric parameters during LS → HS has been shown for several complexes in kHz and THz range [108–110]. At the same time, an opposite effect has been reported for the visible light [111]. This can be associated with the

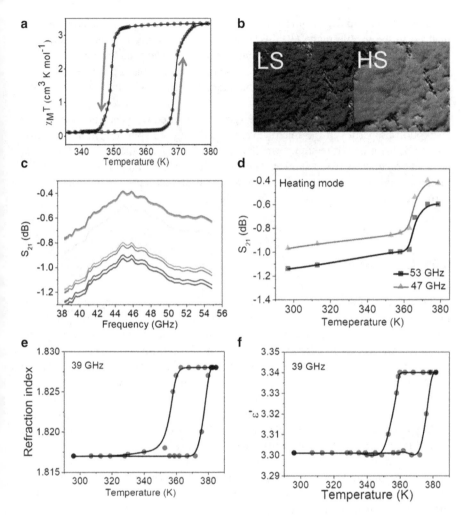

Fig. 6.13 (**a**) Magnetic curve of **2** showing the temperature induced spin transition at 367 K in heating mode and at 349 K in cooling mode; (**b**) photos of the complex in a LS (left) and HS (right) states; (**c**) transmission coefficient (S21) of **2** measured in the 297–379 K temperature range; (**d**) temperature dependence of S_{21} at different frequencies measured in heating mode; temperature dependence of refraction index (**e**) and permittivity (**f**) of **2**

fact, that in the 10^3–10^{12} Hz range the permittivity of a complex is determined by ionic and dipolar polarization. In this case, when the complex changes to the HS state, the increase of temperature and volume facilitates the orientation of dipoles and displacement of ions under the influence of the electric field, which results in the increase of permittivity. Contrastingly, only the atomic type of polarizability is realized in the visible range (10^{14} Hz) and the polarizability is mostly determined by a small deformation of electron clouds around atom/ion nuclei; the increase of temperature and volume leads to the decrease of concentration of polarized particles. As a result, the permittivity in the visible range decreases with the transition to the HS state.

These observations showed that the spin transition complexes can change their absorption properties in the microwave range during the spin transition. This discovery opens a new direction towards application of SCO materials as functional elements of microwave switching devices. Moreover, the huge diversity of SCO materials can provide microwave switches with almost any desired temperature, abruptness, and hysteresis of the spin transition.

6.6   Conclusions

Chiral coordination materials exhibiting spin transition are gaining popularity for many reasons. The chiro-optical response that can be switched during the spin transition can make these materials employed into various photonic devices operating in UV and visible regions. The sensitivity of chiral SCO complexes towards the inclusion of chiral guest molecules opens a unique opportunity for enantioselective detection by means of materials exhibiting spin transition. Moreover, the enantioselective shift of transition temperatures can be potentially detected by any technique suitable for observation of spin transition, e.g. UV-vis, IR, Raman spectroscopy, etc. Different synthetical approaches were successfully used for design and synthesis of chiral SCO compounds. There are many examples of both crystallization induced resolution of enantiomeric SCO entities and synthesis of SCO complexes using ligands or anions with predefined chirality. The second approach allows the production of SCO complexes in higher quantities and different morphological forms, such as nanoparticles, gels, etc.

SCO materials were shown to be capable of switching their microwave absorption ability upon spin transition. The origin of switchable microwave absorption is associated with the change of permittivity in GHz frequency range upon the spin transition. This property of SCO materials allows their implementation into various devices operating in microwave frequency range. Moreover, the variety of SCO materials allows the design of microwave switches with any desired temperature, abruptness and hysteresis of the spin transition.

Acknowledgements Funding for this research was provided by the Ministry of Education and Science of Ukraine grants No. 19BF03-01M, 19BF037-04, 19BF052-04.

References

1. Gütlich P, Goodwin HA (2004) Spin crossover. In: Transition metal compounds I. Springer, Berlin/Heidelberg, pp 1–47
2. Ewald AH, Martin RL, Sinn E et al (1969) Electronic equilibrium between the $^6A_1$ and $^2T_2$ states in iron(III) dithio chelates. Inorg Chem 8:1837–1846
3. Ksenofontov V, Gaspar AB, Gütlich P (2006) Spin crossover. In: Transition metal compounds III. Springer, Berlin/Heidelberg, pp 23–64
4. Qi Y, Müller EW, Spiering H et al (1983) The effect of a magnetic field on the high-spin α low-spin transition in [Fe(phen)$_2$(NCS)$_2$]. Chem Phys Lett 101:503–505
5. Decurtins S, Gütlich P, Köhler CP et al (1984) Light-induced excited spin state trapping in a transition-metal complex: the hexa-1-propyltetrazole-iron (II) tetrafluoroborate spin-crossover system. Chem Phys Lett 105:1–4
6. Ohtani R, Hayami S (2017) Guest-dependent spin-transition behavior of porous coordination polymers. Chem Eur J 23:2236–2248
7. Ni ZP, Liu JL, Hoque MN et al (2017) Recent advances in guest effects on spin-crossover behavior in Hofmann-type metal-organic frameworks. Coord Chem Rev 335:28–43
8. Cambi L, Szegö L (1931) Über die magnetische Susceptibilität der komplexen Verbindungen. Berichte der deutschen chemischen Gesellschaft (A and B Series) 64:2591–2598
9. Rotaru A, Gural'skiy IA, Molnár G et al (2012) Spin state dependence of electrical conductivity of spin crossover materials. Chem Commun 48:4163-4165
10. Hauser A (2004) Spin crossover. In: Gütlich P, Goodwin HA (eds) Transition metal compounds I. Springer, Berlin/Heidelberg, pp 49–58
11. Shepherd HJ, Gural'skiy IA, Quintero CM et al (2013) Molecular actuators driven by cooperative spin-state switching. Nat Commun 4:2607
12. Baker WA Jr, Bobonich HM (1964) Magnetic properties of some high-spin complexes of iron(II). Inorg Chem 3:1184–1188
13. Madeja K, König E (1963) Zur frage der bindungsverhältnise in komplexverbindungen des eisen(II) mit 1, 10-phenanthrolin. J Inorg Nucl Chem 25:377–385
14. Reinen D, Friebel C, Propach V (1974) High-und Low-Spin-Verhalten des Ni^{3+}-Ions in oktaedrischer Koordination.(A) NiF$_6^{3-}$-Polyeder. Z Anorg Allg Chem 408:187–204
15. Garcia Y, Gütlich P (2004) Spin crossover. In: Gütlich P, Goodwin HA (eds) Transition metal compounds II. Springer, Berlin/Heidelberg, pp 49–62
16. Goodwin HA (2004) Spin crossover. In: Gütlich P, Goodwin HA (eds) Transition metal compounds II. Springer, Berlin/Heidelberg, pp 23–47
17. Bethe H (1929) Termaufspaltung in kristallen. Ann Phys 395:133–208
18. Roubeau O (2012) Triazole-based one-dimensional spin-crossover coordination polymers. Chem Eur J 18:15230–15244
19. Niel V, Martinez-Agudo JM, Munoz MC et al (2001) Cooperative spin crossover behavior in cyanide-bridged Fe(II)–M(II) bimetallic 3D Hofmann-like networks (M= Ni, Pd, and Pt). Inorg Chem 40:3838–3839
20. Tweedle MF, Wilson LJ (1998) Variable spin iron(III) chelates with hexadentate ligands derived from triethylenetetramine and various salicylaldehydes. Synthesis, characterization, and solution state studies of a new $^2T \rightleftharpoons {}^6A$ spin equilibrium system. J Am Chem Soc 98:4824–4834
21. Kroeber J, Audiere JP, Claude R et al (1994) Spin transitions and thermal hysteresis in the molecular-based materials [Fe(Htrz)$_2$(trz)](BF$_4$) and [Fe(Htrz)$_3$](BF$_4$)$_2$·H$_2$O (Htrz=1,2,4-4H-triazole; trz=1,2,4-triazolato). Chem Mater 6:1404–1412
22. Lavrenova LG, Shakirova OG (2013) Spin crossover and thermochromism of iron(II) coordination compounds with 1,2,4-Triazoles and Tris(pyrazol-1-yl)methanes. Eur J Inorg Chem 2013:670–682
23. Peng H, Molnár G, Salmon L et al (2015) Matrix-free synthesis of spin crossover micro-rods showing a large hysteresis loop centered at room temperature. Chem Commun 51:9346–9349

24. Pittala N, Thétiot F, Triki S et al (2017) Cooperative 1D triazole-based spin crossover FeII material with exceptional mechanical resilience. Chem Mater 29:490–494
25. Grosjean A, Daro N, Kauffmann B et al (2011) The 1-D polymeric structure of the [Fe (NH$_2$trz)$_3$](NO$_3$)$_2$·nH$_2$O (with n= 2) spin crossover compound proven by single crystal investigations. Chem Commun 47:12382–12384
26. Koningsbruggen PJ (2004) Spin crossover. In: Gütlich P, Goodwin HA (eds) Transition metal compounds I. Springer, Berlin/Heidelberg, pp 123–149
27. Hofmann KA, Küspert F (1897) Verbindungen von kohlenwasserstoffen mit metallsalzen. Z Anorg Chem 15:204–207
28. Kitazawa T, Gomi Y, Takahashi M et al (1996) Spin-crossover behaviour of the coordination polymer Fe^{II}(C$_5$H$_5$N)$_2$Ni^{II}(CN)$_4$. J Mater Chem 6:119–121
29. Piñeiro-López L, Valverde-Muñoz FJ, Seredyuk M et al (2017) Guest induced strong cooperative one-and two-step spin transitions in highly porous iron(II) Hofmann-type metal–organic frameworks. Inorg Chem 56:7038–7047
30. Piñeiro-López L, Valverde-Muñoz FJ, Seredyuk M et al (2018) Cyanido-bridged Fe^{II}–M^I dimetallic Hofmann-like spin-crossover coordination polymers based on 2,6-naphthyridine. Eur J Inorg Chem 2018:289–296
31. Kosone T, Tomori I, Kanadani C et al (2010) Unprecedented three-step spin-crossover transition in new 2-dimensional coordination polymer {Fe^{II}(4-methylpyridine)$_2$[Au^I (CN)$_2$]$_2$}. Dalton Trans 39:1719–1721
32. Wei RM, Kong M, Cao F et al (2016) Water induced spin-crossover behaviour and magneto-structural correlation in octacyanotungstate(IV)-based iron(II) complexes. Dalton Trans 45:18643–18652
33. Gural'skiy IA, Shylin SI, Ksenofontov V et al (2019) Pyridazine-supported polymeric cyanometallates with spin transitions. Eur J Inorg Chem 2019:4532–4537
34. Zhang CJ, Lian KT, Huang GZ et al (2019) Hysteretic four-step spin-crossover in a 3D Hofmann-type metal–organic framework with aromatic guest. Chem Commun 55:11033–11036
35. Sciortino NF, Scherl-Gruenwald KR, Chastanet G et al (2012) Hysteretic three-step spin crossover in a thermo-and photochromic 3D pillared Hofmann-type metal–organic framework. Angew Chem Int Ed 51:10154–10158
36. Rodríguez-Velamazán JA, Carbonera C, Castro M et al (2010) Two-step thermal spin transition and LIESST relaxation of the polymeric spin-crossover compounds Fe(X-py)$_2$[Ag(CN)$_2$]$_2$ (X= H, 3-methyl, 4-methyl, 3, 4-dimethyl, 3-Cl). Chem Eur J 16:8785–8796
37. Hiiuk VM, Shova S, Rotaru A et al (2019) Room temperature hysteretic spin crossover in a new cyanoheterometallic framework. Chem Commun 55:3359–3362
38. Munoz MC, Real JA (2011) Thermo-, piezo-, photo-and chemo-switchable spin crossover iron(II)-metallocyanate based coordination polymers. Coord Chem Rev 255:2068–2093
39. Long GJ, Grandjean F, Reger DL (2004) Spin crossover. In: Gütlich P, Goodwin HA (eds) Transition metal compounds I. Springer, Berlin/Heidelberg, pp 91–122
40. Shalabaeva V, Rat S, Manrique-Juarez MD et al (2017) Vacuum deposition of high-quality thin films displaying spin transition near room temperature. J Mater Chem C 5:4419–4425
41. Real JA, Andres E, Munoz MC et al (1995) Spin crossover in a catenane supramolecular system. Science 268:265–267
42. Guionneau P, Marchivie M, Bravic G et al (2004) Spin crossover. In: Gütlich P, Goodwin HA (eds) Transition metal compounds II. Springer, Berlin/Heidelberg, pp 97–128
43. Halder GJ, Kepert CJ, Moubaraki B et al (2002) Guest-dependent spin crossover in a nanoporous molecular framework material. Science 298:1762–1765
44. Ohba M, Yoneda K, Agustí G et al (2009) Bidirectional chemo-switching of spin state in a microporous framework. Angew Chem Int Ed 48:4767–4771
45. Quintero CM, Costa JS, Demont P et al (2014) Spin crossover composite materials for electrothermomechanical actuators. J Mater Chem C 2:2949–2955

46. Salmon L, Molnár G, Zitouni D et al (2010) A novel approach for fluorescent thermometry and thermal imaging purposes using spin crossover nanoparticles. J Mater Chem 20:5499–5503

47. Nagy V, Halász K, Carayon MT et al (2014) Cellulose fiber nanocomposites displaying spin-crossover properties. Colloids and surfaces. A, Physicochemical and engineering aspects. 456:35–40

48. Kahn O, Martinez CJ (1998) Spin-transition polymers: from molecular materials toward memory devices. Science 279:44–48

49. Halcrow MA (2016) The effect of ligand design on metal ion spin state—lessons from spin crossover complexes. Crystals 6(5):58

50. Scott HS, Staniland RW, Kruger PE (2018) Spin crossover in homoleptic Fe(II) imidazolylimine complexes. Coord Chem Rev 362:24–43

51. Attwood M, Turner SS (2017) Back to back 2,6-bis(pyrazol-1-yl)pyridine and 2,2':6',2"-terpyridine ligands: untapped potential for spin crossover research and beyond. Coord Chem Rev 353:247–277

52. Feltham HL, Barltrop AS, Brooker S (2017) Spin crossover in iron(II) complexes of 3,4,5-tri-substituted-1,2,4-triazole (Rdpt), 3,5-di-substituted-1,2,4-triazolate (dpt⁻), and related ligands. Coord Chem Rev 344:26–53

53. Craig GA, Roubeau O, Aromí G (2014) Spin state switching in 2,6-bis(pyrazol-3-yl)pyridine (3-bpp) based Fe(II) complexes. Coord Chem Rev 269:13–31

54. Cirera J (2014) Guest effect on spin-crossover frameworks. Rev Inorg Chem 34:199–216

55. Quintero CM, Félix G, Suleimanov I et al (2014) Hybrid spin-crossover nanostructures. Beilstein J Nanotechnol 5:2230–2239

56. Kumar KS, Ruben M (2017) Emerging trends in spin crossover (SCO) based functional materials and devices. Coord Chem Rev 346:176–205

57. Li H, Peng H (2018) Recent advances in self-assembly of spin crossover materials and their applications. Curr Opin Colloid Interface Sci 35:9–16

58. Brooker S (2015) Spin crossover with thermal hysteresis: practicalities and lessons learnt. Chem Soc Rev 44:2880–2892

59. Molnár G, Mikolasek M, Ridier K et al (2019) Molecular Spin Crossover Materials: Review of the Lattice Dynamical Properties. Ann Phys 531:1900076

60. Unruh D, Homenya P, Kumar M et al (2016) Spin state switching of metal complexes by visible light or hard X-rays. Dalton Trans 45:14008–14018

61. Shatruk M, Phan H, Chrisostomo BA et al (2015) Symmetry-breaking structural phase transitions in spin crossover complexes. Coord Chem Rev 289:62–73

62. Guionneau P (2014) Crystallography and spin-crossover. A view of breathing materials. Dalton Trans 43:382–393

63. Halcrow MA (2014) Spin-crossover compounds with wide thermal hysteresis. Chem Lett 43:1178–1188

64. Zahn S, Canary JW (2000) Electron-induced inversion of helical chirality in copper complexes of N, N-dialkylmethionines. Science 288:1404–1407

65. Crassous J (2009) Chiral transfer in coordination complexes: towards molecular materials. Chem Soc Rev 38:830–845

66. Murguly E, Norsten TB, Branda NR (2001) Nondestructive data processing based on chiroptical 1,2-Dithienylethene photochromes. Angew Chem Int Ed 40:1752–1755

67. Matsukizono H, Kuroiwa K, Kimizuka N (2008) Lipid-packaged linear iron(II) triazole complexes in solution: controlled spin conversion via solvophobic self-assembly. J Am Chem Soc 130:5622–5623

68. Sunatsuki Y, Ikuta Y, Matsumoto N et al (2003) An unprecedented homochiral mixed-valence spin-crossover compound. Angew Chem Int Ed 42:1614–1618

69. Hashibe T, Fujinami T, Furusho D et al (2011) Chiral spin crossover iron (II) complex, fac-Λ-[Fe^{II}(HL^R)$_3$](ClO$_4$)$_2$·EtOH (HL^R= 2-methylimidazol-4-yl-methylideneamino-R-(+)-1-methylphenyl). Inorg Chim Acta 375:338–342

70. Ould Hamouda A, Dutin F, Degert J et al (2019) Study of the photoswitching of a Fe(II) chiral complex through linear and nonlinear ultrafast spectroscopy. J Phys Chem Lett 10:5975–5982
71. Gao WQ, Meng YS, Liu CH et al (2019) Spin crossover and structural phase transition in homochiral and heterochiral $Fe[(pybox)_2]^{2+}$ complexes. Dalton Trans 48:6323–6327
72. Naim A, Bouhadja Y, Cortijo M et al (2018) Design and study of structural linear and nonlinear optical properties of Chiral $[Fe(phen)_3]^{2+}$ complexes. Inorg Chem 57:14501–14512
73. Sunatsuki Y, Ohta H, Kojima M et al (2004) Supramolecular spin-crossover iron complexes based on imidazole– imidazolate hydrogen bonds. Inorg Chem 43:4154–4171
74. Ohkoshi SI, Takano S, Imoto K et al (2014) 90-degree optical switching of output second-harmonic light in chiral photomagnet. Nat Photonics 8:65–71
75. Sekimoto Y, Karim MR, Saigo N et al (2017) Crystal structures and spin-crossover behavior of iron(II) complexes with chiral and racemic ligands. Eur J Inorg Chem 2017:1049–1053
76. Ru J, Yu F, Shi PP et al (2017) Three properties in one coordination complex: chirality, spin crossover, and dielectric switching. Eur J Inorg Chem 2017:3144–3149
77. Burrows KE, McGrath SE, Kulmaczewski R et al (2017) Spin states of homochiral and heterochiral isomers of $[Fe(PyBox)_2]^{2+}$ derivatives. Chem Eur J 23:9067–9075
78. Ren DH, Qiu D, Pang CY et al (2015) Chiral tetrahedral iron(II) cages: diastereoselective subcomponent self-assembly, structure interconversion and spin-crossover properties. Chem Commun 51:788–791
79. Tian L, Pang CY, Zhang FL et al (2015) Toward chiral iron(II) spin-crossover grafted resins with imidazole Schiff-base ligands. Inorg Chem Commun 53:55–59
80. Qin LF, Pang CY, Han WK et al (2015) Optical recognition of alkyl nitrile by a homochiral iron(II) spin crossover host. Cryst Eng Commun 17:7956–7963
81. Ren DH, Sun XL, Gu L et al (2015) A family of homochiral spin-crossover iron(II) imidazole Schiff-base complexes. Inorg Chem Commun 51:50–54
82. Wang Q, Venneri S, Zarrabi N et al (2015) Stereochemistry for engineering spin crossover: structures and magnetic properties of a homochiral vs. racemic $[Fe(N_3O_2)(CN)_2]$ complex. Dalton Trans 44:6711–6714
83. Gu ZG, Pang CY, Qiu D et al (2013) Homochiral iron(II) complexes based on imidazole Schiff-base ligands: syntheses, structures, and spin-crossover properties. Inorg Chem Commun 35:164–168
84. Reshetnikov VA, Szebesczyk A, Gumienna-Kontecka E et al (2015) Chiral spin crossover nanoparticles and gels with switchable circular dichroism. J Mater Chem C 3:4737–4741
85. Kolb JS, Thomson MD, Novosel M et al (2007) Characterization of Fe(II) complexes exhibiting the ligand-driven light-induced spin-change effect using SQUID and magnetic circular dichroism. C R Chim 10:125–136
86. Han WK, Qin LF, Pang CY et al (2017) Polymorphism of a chiral iron(II) complex: spin-crossover and ferroelectric properties. Dalton Trans 46:8004–8008
87. Bartual-Murgui C, Pineiro-Lopez L, Valverde-Muñoz FJ et al (2017) Chiral and racemic spin crossover polymorphs in a family of mononuclear iron(II) compounds. Inorg Chem 56:13535–13546
88. Qin LF, Pang CY, Han WK et al (2016) Spin crossover properties of enantiomers, co-enantiomers, racemates, and co-racemates. Dalton Trans 45:7340–7348
89. Romero-Morcillo T, Valverde-Muñoz FJ, Muñoz MC et al (2015) Two-step spin crossover behaviour in the chiral one-dimensional coordination polymer $[Fe(HAT)(NCS)_2]_\infty$. RSC Adv 5:69782–69789
90. Liu W, Bao X, Mao LL et al (2014) A chiral spin crossover metal–organic framework. Chem Commun 50:4059–4061
91. Sunatsuki Y, Miyahara S, Sasaki Y et al (2012) Conglomerate crystallization, chiral recognition and spin-crossover in a host–guest complex consisting of Fe^{III} complexes (host) and $[Cr(ox)_3]^{3-}$ (guest). Cryst Eng Commun 14:6377–6380
92. Sato T, Iijima S, Kojima M et al (2009) Assembling into chiral crystal of spin crossover iron(II) complex. Chem Lett 38:178–179

93. Sato T, Nishi K, Iijima S et al (2009) One-step and two-step spin-crossover iron(II) complexes of ((2-methylimidazol-4-yl)methylidene)histamine. Inorg Chem 48:7211–7229
94. Brewer CT, Brewer G, Butcher RJ et al (2007) Synthesis and characterization of a spin crossover iron(II)–iron(III) mixed valence supramolecular pseudo-dimer exhibiting chiral recognition, hydrogen bonding, and π–π interactions. Dalton Trans 3:295–298
95. Hoshino N, Iijima F, Newton GN et al (2012) Three-way switching in a cyanide-bridged [CoFe] chain. Nat Chem 4:921
96. Jakobsen VB, O'Brien L, Novitchi G et al (2019) Chiral resolution of a Mn^{3+} spin crossover complex. Eur J Inorg Chem 2019:4405–4411
97. Gómez V, Saenz de Pipaon C, Maldonado-Illescas P et al (2015) Easy excited-state trapping and record high T_{TIESST} in a spin-crossover polyanionic Fe^{II} trimer. J Am Chem Soc 137:11924–11927
98. Surma MA, Szczepaniak A, Króliczewski J (2014) Comparative studies on detergent-assisted apocytochrome b_6 reconstitution into liposomal bilayers monitored by Zetasizer instruments. PLoS One 9:e111341
99. Gural'skiy IA, Kucheriv OI, Shylin SI et al (2015) Enantioselective guest effect on the spin state of a chiral coordination framework. Chem Eur J 21:18076–18079
100. El-Hinnawy N, Borodulin P, Wagner BP et al (2014) Low-loss latching microwave switch using thermally pulsed non-volatile chalcogenide phase change materials. Appl Phys Lett 105:013501
101. Young RM, El-Hinnawy N, Borodulin P et al (2014) Thermal analysis of an indirectly heat pulsed non-volatile phase change material microwave switch. J Appl Phys 116:054504
102. Singh T, Mansour RR (2018) Chalcogenide Phase Change Material GeTe Based Inline RF SPST Series and Shunt Switches. IEEE MTT-S Int. Microw. Workshop Ser. Adv. Mater. Processes RF THz Appl. (IMWS-AMP) 1:1–3
103. Yang S, Vaseem M, Shamim A (2019) Fully inkjet-printed VO_2-based radio-frequency switches for flexible reconfigurable components. Adv Mater Technol 4:1800276
104. Dumas-Bouchiat F, Champeaux C, Catherinot A, Crunteanu A, Blondy P (2007) Rf-microwave switches based on reversible semiconductor-metal transition of VO_2 thin films synthesized by pulsed-laser deposition. Appl Phys Lett 91:223505
105. Kucheriv OI, Oliynyk VV, Zagorodnii VV et al (2016) Spin-crossover materials towards microwave radiation switches. Sci Rep 6:38334
106. Gural'skiy IA, Golub BO, Shylin SI et al (2016) Cooperative high-temperature spin crossover accompanied by a highly anisotropic structural distortion. Eur J Inorg Chem 2016:3191–3195
107. Gural'skiy I A, Kucheriv OI, Oliynyk VV et al (2019) Ukraine patent No 139195
108. Bousseksou A, Molnár G, Demont P et al (2003) Observation of a thermal hysteresis loop in the dielectric constant of spin crossover complexes: towards molecular memory devices. J Mater Chem 13:2069–2071
109. Mounaix P, Lascoux N, Degert J et al (2005) Dielectric characterization of $[Fe(NH_2-trz)_3]$ $Br_2 \cdot H_2O$ thermal spin crossover compound by terahertz time domain spectroscopy. Appl Phys Lett 87:244103
110. Viquerat B, Degert J, Tondusson M et al (2011) Time-domain terahertz spectroscopy of spin state transition in $[Fe(NH_2-trz)_3]^{2+}$ spin crossover compounds. Appl Phys Lett 99:061908
111. Loutete-Dangui ED, Varret F, Codjovi E et al (2007) Thermal spin transition in $[Fe(NH_2-trz)_3]Br_2$ investigated by spectroscopic ellipsometry. Phys Rev B 75:184425

Chapter 7
Grain Boundary Diffusion Dominated Mixing and Solid State Reactions in Magnetic Thin Films

Gábor L. Katona and Szilvia Gulyas

Abstract Magnetic thin films with $L1_0$ ordered structure usually exhibit high magnetocrystalline anisotropy, reasonable coercivity, saturation magnetization, excellent corrosion resistance. Examples for such systems are Fe-Pt, Co-Pt, Fe-Pd, Fe-Ni. The ordered phases are almost exclusively fabricated by employing high temperature annealing or deposition on heated substrates, because of the activation of the necessary diffusion processes. At lower temperatures bulk diffusion slows down and becomes practically negligible, while diffusion along defects, like grain boundaries and dislocations is still available. During grain boundary interdiffusion the migration of the boundary can happen, and the migrating boundary leaves an alloyed zone behind. This phenomenon is called diffusion induced grain boundary migration (DIGM), and can lead to even full intermixing if the grain size is small enough. Here, after a short introduction of this process, examples of ordered phase formation at low temperatures in magnetic thin films will be reviewed.

Keywords Grain boundary diffusion · DIGM · Solid state reaction · Magnetic thin film · Ordered phase formation

G. L. Katona (✉)
Faculty of Science and Technology, Department of Solid State Physics, University of Debrecen, Debrecen, Hungary
e-mail: katonag@science.unideb.hu

S. Gulyas
Faculty of Science and Technology, Department of Solid State Physics, University of Debrecen, Debrecen, Hungary

Doctoral School of Physics, University of Debrecen, Debrecen, Hungary

© Springer Nature B.V. 2020
A. Kaidatzis et al. (eds.), *Modern Magnetic and Spintronic Materials*, NATO Science for Peace and Security Series B: Physics and Biophysics, https://doi.org/10.1007/978-94-024-2034-0_7

7.1 Introduction

Depending on temperature, the speed of diffusion not only changes because of the temperature dependence of the diffusion coefficient, but also because of different diffusion paths. In case of bulk (lattice) diffusion the mixing of atoms happens in the matrix, inside grains if the material is polycrystalline. If there are defects, like grain boundaries, dislocations, these act as diffusion short circuits, because the diffusivity is much higher in these regions as compared to the lattice. These high diffusivity paths become extremely important when the lattice diffusivity slows down such that it becomes negligible in the investigated time frame. This usually occurs if the temperature drops below approximately half of the melting point. At these temperatures the bulk diffusion coefficient is so low, that the diffusion length ($\sqrt{Dt}$) is usually well below the width of the grain boundary, thus effective diffusion can only occur in grain boundaries, dislocations. In such case, one usually does not expect any extended structural or phase transformation, since atomic migration is restricted to the grain boundaries.

However, there is a special phenomenon called Diffusion Induced Grain boundary Migration (DIGM), when the interdiffusion of two components along grain boundaries induce the shift of the boundaries, which leaves behind an alloyed region during its motion. This shift of the boundary can at least partly alloy the materials. DIGM was discovered by den Broeder [9]. A classical illustration is show in Fig. 7.1 where the zincification of an iron foil can be seen [19]. An important feature of DIGM is that the region left behind by the boundary has different composition than that of the grain. This phenomenon was observed in several metallic system [1, 9, 19], in semiconductors [40] and also in oxides [4, 35, 50]. The alloying caused by DIGM can be applied in practical applications, see e.g [3, 10, 14, 16–18, 24, 38, 44].

The explanation of this phenomenon is that during grain boundary interdiffusion usually there is a net volume flow because the volume of the diffusing species and

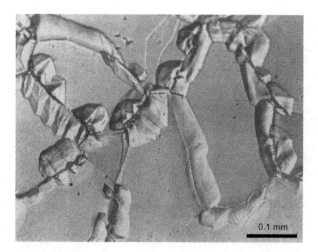

Fig. 7.1 Optical micrograph illustrating the surface relief that accompanies the zincification of an iron foil, heated in Zn vapor for 4 h at 600 °C. Interference contrast. (Reproduced from [19], with permission from Elsevier)

also their diffusion coefficient differs, similarly to the bulk case [2, 41]. This net volume flow creates diffusion, in this case grain boundary diffusion, induced strain that causes stress. This stress can stop the diffusion since the accumulated stress gradient acts as an additional driving force [26, 37]. The other possibility is the relaxation of this stress [1]. This can happen by the shift of the boundary, which is the already mentioned DIGM, when the moving boundary leaves behind an alloyed region by which the induced strain decreases. Additionally Diffusion Induced Recrystallization (DIR) was also observed [31, 32], when the accumulated stress relaxes by nucleation of new grains with composition different from that of the matrix. Regarding intermetallic compounds it was shown shortly after the discovery of DIGM, that in the Ag-Pd system at low temperatures a compound phase formed [43], although it was not clear that time whether this formation is based on DIGM.

In the following, several examples of low temperature ordered phase formation will be presented focusing on magnetic thin films, especially the Fe-Pt system. This system is important for e.g. magnetic storage, spintronic applications, since $L1_0$-FePt thin films exhibit high magnetic anisotropy, high coercivity which is needed for these applications [5, 28, 30, 33, 48]. The main challenge is forming the desired $L1_0$ phase, because after deposition the films exhibit a disordered A1 structure.

Selecting a suitable substrate and growing the film at high temperature results in formation of the ordered phase [7, 11, 39, 47]. Alternatively, annealing of a disordered A1-FePt phase at high temperature (see e.g. [49]) also forms the ordered structure. However, the temperature needed for these processes are quite high, thus lowering of the order-disorder transition temperature is needed. For applications, the control of crystal orientation (texture) and grain size is also necessary. These difficulties led to the development of several other methods. The most investigated processes utilize the addition of third elements, like Cu [6, 15, 27, 29, 52], Au [12, 34] or Ag [34, 42, 46, 51, 53]. These additions have different roles. Cu forms a ternary FePtCu alloy [6, 21, 29], with the Cu substituting the Fe in the ordered structure, while Ag segregates from the FePt phase [8, 25]. Au behaves similarly to Ag, i.e. it does not sit in the FePt lattice [13]. The substitution in case of Cu or the segregation (and the corresponding diffusion) in case of Au and Ag cause stress in the film which is supposed to be the main reason of the change of ordering kinetics [13, 20]. However, even with these methods the temperatures reported for the formation of the ordered phase are relatively high.

It will be briefly reviewed in the following, that utilizing grain boundary diffusion related processes like DIGM and accompanying compound formation at temperatures below 400 °C the ordered phase appears and even full transformation of the film can be expected.

7.2 Fe/Pt System

Considering pure Fe/Pt films, without any additional component, Katona et al. [22] fabricated bilayers of 15 nm Fe and 15 nm Pt by dc magnetron sputtering on SiO_2 substrate. The as received Pt/Fe/SiO_2 samples were then annealed in vacuum at 340 °C. The samples were analyzed by Secondary Neutral Mass Spectrometry

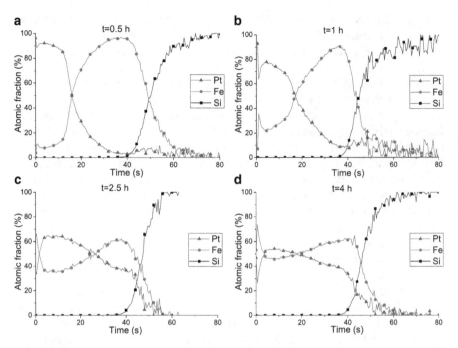

Fig. 7.2 Composition profiles of Pt(15 nm)/Fe(15 nm)/SiO2 samples annealed at 340 °C for different times of (**a**) 0.5 h, (**b**) 1 h, (**c**) 2.5 h, and (**d**) 4 h. (Reproduced from [22], with permission from Springer)

(SNMS) depth profiling, X-Ray Diffraction (XRD) and Transmission Electron Microscopy (TEM). Fig. 7.2 shows the depth profiles of the as deposited sample and after annealing at 340 °C for 0.5, 1, 2.5 and 4 h. As it can be seen, the layers intermix even at this relatively low temperature, where the bulk diffusion length (corresponding to diffusion in the grains) for the investigated annealing times is about 10^{-12} m, meaning negligible diffusion. The intermixing is thus a consequence of grain boundary diffusion. However, even if we assume that grain boundaries are completely filled up by the diffusing component it cannot explain the high concentration of Fe in the Pt layer after 1 h, and the high concentration of both components in the other layer after further annealing. The final state after 4 h annealing is almost complete homogenization resulting in a FePt layer with approx. 50/50 at% composition. The XRD pattern of the sample annealed for 4 h is shown in Fig. 7.3. The peaks are identified as reflections from disordered (A1) and ordered (L1$_0$) FePt phases, proving that the mixing of Fe and Pt resulted in partly ordered film.

Shamis et al. [36] also investigated dc magnetron sputtered Pt(15 nm)/Fe(15 nm)/ substrate samples via in/situ electrical resistivity measurements during annealing accompanied by depth profiling, XRD and magnetic measurements. The substrate was MgO in case of electrical resistivity measurements and SiO$_2$ in case of the other methods. During measurements the samples were heated up with a heating rate of 0.5 °C/s to 620 °C and the resistivity of the sample was measured via conventional 4-point setup. SNMS depth profiles, XRD patterns, magnetic parameters were

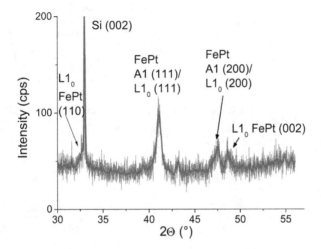

Fig. 7.3 XRD Θ–2Θ pattern of the Pt(15 nm)/Fe(15 nm)/SiO2 /Si sample post-annealed for 4 h at 340 °C. The smoothed scan is shown with a bold line (blue online) for convenience, whereas the raw scan can be seen in grey. (Reproduced from [22], with permission from Springer)

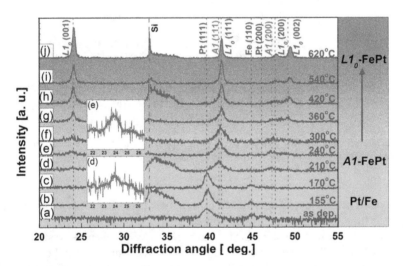

Fig. 7.4 XRD (θ − 2θ) scans of Pt/Fe films (**a**) after deposition and (**b**)-(**j**) post-annealing at different temperatures. Insets show the appearance of the L1$_0$(0 0 1) superstructure reflection. (Reproduced from [36], with permission from IOP Publishing)

measured by stopping at selected temperatures. The resistivity-temperature curves revealed four regions of the transformation. The regions were interpreted based on the accompanying measurements. Fig. 7.4 shows the XRD pattern, Fig. 7.5 presents the depth profiles of the samples at the selected temperatures. As it can be seen from the depth profiles, even after reaching 210 °C (Fig. 7.5c) the layers are already intermixed. Similarly to the previous investigation, the composition of the

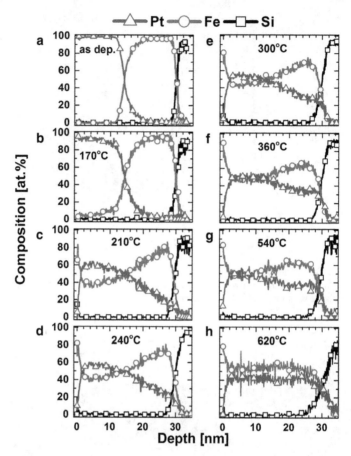

Fig. 7.5 Composition versus depth profiles of Pt/Fe bilayers (**a**) after deposition and (**b**)-(**h**) post-annealing at different temperatures. (Reproduced from [36], with permission from IOP Publishing)

components in the other layer is higher, especially in case of the Pt layer, than can be expected from the filling up of grain boundaries.

This is confirmed by the XRD pattern (Fig. 7.4d) where formation of the disordered A1-FePt phase could be seen while the individual Pt and Fe reflections vanished. Examining carefully the XRD pattern even a small peak associated with the ordered $L1_0$ structure could be identified. Magnetic measurements also confirmed the appearance of the ordered structure (for the M-H loops see Fig. 7.3 in [36]). The intermixing of the layers continued with the increase of the final temperature (Fig. 7.5e) together with the small increase of the ordered phase fraction as it can be seen from the XRD patterns after 240 °C and 300 °C. Above 300 °C the reflections corresponding to the $L1_0$ ordered phase drastically strengthened indicating the increase of the amount of the phase (Fig. 7.4g). The depth profile obtained after annealing up to 360 °C show advanced intermixing with an approximately equiatomic composition in place of the original Pt layer. Further increase of the maximum temperature finally results in a homogenous atomic distribution in the whole thickness (Fig. 7.5h) at 620 °C. The corresponding XRD pattern (Fig. 7.4j) shows a practically fully ordered FePt phase, which is not textured, but it should be noted that even the sample annealed to 420 °C is highly ordered (Fig. 7.4h). The transformation of the whole bilayer into ordered FePt phase was also confirmed by magnetic measurements, where the sample annealed up to 620 °C was annealed for a second cycle and was compared to a cycle measured on sample annealed at high temperature, thus having fully ordered structure. The comparison proved that the samples behave exactly the same during the R(T) measurement.

The interpretation in both cases rely on the fact, that bulk diffusion in the grains itself is negligible. First, there is mutual grain boundary diffusion and filling up of both components in the other layer. The high concentration of the components in the other layer cannot be explained by this filling up, but with the process of DIGM and the following reaction layer formation (GBDIREAC). During this process, the stress accumulated because of the filling of the grain boundaries relaxes by the perpendicular shift of the boundaries. The moving grain boundaries leave behind an alloyed region, which becomes ordered, because in this system an intermetallic phase exists. This in accordance with the depth profiles, where a gradual increase of the concentration can be seen. It should be taken into account, that the depth profile gives an average composition over a relatively large area, thus it is the average of alloyed and unalloyed regions. As the grain boundaries sweep, the left behind reaction layers (the alloyed zone) becomes larger and larger increasing the average composition. Since the grain size is small, the shift of the boundaries finally result in complete transformation of the originally pure layers. This is also in accordance with the fact, that no conventional reaction layer formed and grow at the interface.

7.3 Fe/Ag/Pt

Low temperature investigations were also carried out with Ag as a third component. Katona et al. [22] also prepared Pt(15 nm)/Ag(10 nm)/Fe(15 nm)/SiO$_2$ samples and annealed the same way as the bilayer in the previous section. The depth profile obtained after 1 h at 340 °C (Fig. 7.6b) shows intensive intermixing between the Ag and Pt layers together with some Pt penetration into the Fe layer. Increasing the annealing time the mixing of Ag and Pt continues together with Pt penetration to the Fe layer (Fig. 7.6c, d). Finally a close to homogeneous layer forms with considerable amount of Ag present at the free surface, suggesting a bilayer structure (Fig. 7.6e). The XRD pattern of the sample annealed for 4 h is shown in Fig. 7.7. The reflections correspond to FePt and Ag. The FePt phase is at least partially ordered as it can be seen from the appearance of the two reflections around 33° and 49°, which originates purely from the L1$_0$ ordered structure. The presence of Ag reflection together with the depth profile indicates that there is a significant Ag layer at the free surface, proving that finally a bilayer structure is formed. Looking into the details of the

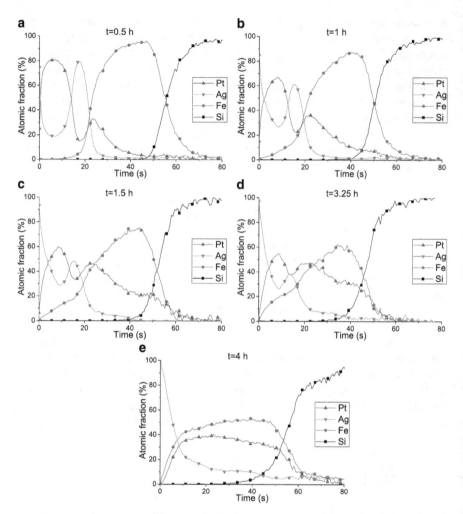

Fig. 7.6 Time evolution of the composition profiles of Pt(15 nm)/Ag(10 nm)/Fe(15 nm)/SiO₂/Si samples post-annealed at 340 °C for different annealing times of (**a**) 0.5 h, (**b**) 1 h, (**c**) 1.5, (**d**) 3.25 h, and (**e**) 4 h. (Reproduced from [22], with permission from Springer)

phase formation kinetics it was found, that first a metastable AgPt phase has formed by diffusion of Pt and Ag into the grain boundaries of the Ag and Pt layers, respectively. Afterwards this metastable AgPt phase is decomposed as the Pt leaves this phase in favour of forming FePt, by diffusing into the Fe layer. It should be emphasized, that neither the AgPt nor the FePt layer was formed by the usual reaction layer formation starting from the original interfaces, but are formed simultaneously in the whole layer thickness. This was interpreted by DIGM and GBDIREAC, first between Ag and Pt, and afterwards in the Fe layer.

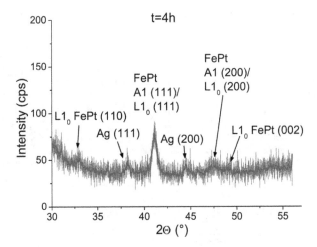

Fig. 7.7 XRD Θ–2Θ patterns of the Pt(15 nm)/Ag(10 nm)/Fe(15 nm)/SiO$_2$/Si sample post-annealed at 340 °C for 4 h. The smoothed scan is shown with a bold line (blue online) for convenience, whereas the raw scan can be seen in grey. (Reproduced from [22], with permission from Springer)

The reverse system of Fe/Ag/Pt/SiO$_2$ was investigated by Katona et al. [23] with the same thicknesses as in the previous case; 15 nm Fe, 10 nm Ag and 15 nm Pt. The film was prepared by dc magnetron sputtering. The samples were annealed between 245 and 390 °C for various duration ranging from 1 to 52 h and analyzed by SNMS depth profiling, XRD and magnetic measurements (SQUID-VSM). At 300 °C after 1 h the depth profile shows a reasonable amount of Ag in the Pt layer together with Pt penetration into the Ag layer (Fig. 7.8a). This advances after 2.5 h annealing at this temperature to approximately 30% Pt in the Ag layer and about 20% Ag in the Pt layer (Fig. 7.8c). The Fe layer is practically intact during these annealings. Increasing the temperature to 340 °C results in a more advanced intermixing between Ag and Pt after 1 h (Fig. 7.9a), suggesting formation of an AgPt phase. At this temperature there is already some Pt appearing in the Fe layer, however FePt compound layer was not detected at the Fe/AgPt contact plane. Longer annealing leads to advanced penetration of Pt into the Fe layer while depleting the previously formed AgPt phase (Fig. 7.8b, c). As a consequence, an Ag layer appears at the substrate just as it would have 'moved' to the substrate. The apparent enrichment of Fe at the surface was found to be an artifact because of the oxidation of the surface region of Fe. In order to approach the final structure annealing at 390 °C was also carried out which resulted in an almost equiatomic FePt layer at the surface and a Ag layer at the substrate (Fig. 7.10). The latter is not pure but contains also Fe and Pt which was supposed to be in a form of FePt precipitates. Corresponding XRD patterns can be seen in Fig. 7.11. In this case formation of ordered FePt is much less visible in the XRD patterns, however week reflections from the ordered phase appear after 52 h at 340 °C. Important finding was the clear reflection from Ag in the

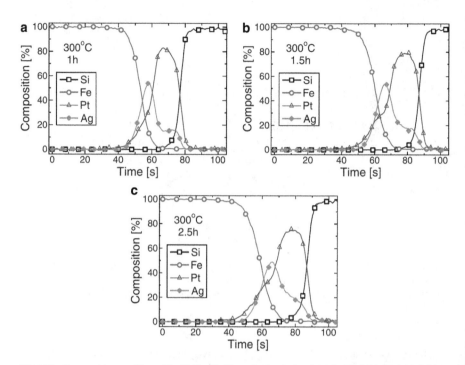

Fig. 7.8 Composition profiles of the Fe/Ag/Pt films obtained after (**a**) 1 h, (**b**) 1.5 h, and (**c**) 2.5 h at 300 °C. The symbols are for identification of the curves. (Reproduced from [23], with permission from IOP Publishing)

XRD pattern. M-H hysteresis loops were also recorded and these proved the appearance of ordered FePt phase, as the coercivity increased to 10 kOe after 52 h at 340 °C. The low temperature results were compared to high temperature (600–900 °C) flash (30s) annealing and it was found, that the final state corresponded to annealing above 700 °C but below 800 °C.

The process of the low temperature annealing is pictured schematically in Fig. 7.12. First the Ag and Pt layers intermix with more advanced diffusion in the Ag layer. Afterwards a AgPt phase forms, while Pt diffusion into the Fe layer advances. Finally there is a FePt phase closer to the surface and a Ag layer with FePt around the grain boundaries at the substrate.

Comparing the two systems with Ag interlayer it was concluded, that regardless of the original structure, the final state is a FePt and a Ag layer, the latter appearing at the place of the Pt layer in the starting sequence. It was suggested in Katona et al. [23], which even for other structures, like Pt/Ag/Fe/Ag/Pt the result would be similar, in this example as Ag/FePt/Ag.

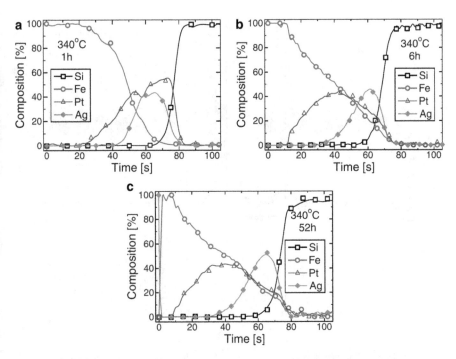

Fig. 7.9 Composition profiles of the Fe/Ag/Pt films obtained after (**a**) 1 h, (**b**) 6 h, and (**c**) 52 h at 340 °C. The symbols are for identification of the curves. (Reproduced from [23], with permission from IOP Publishing)

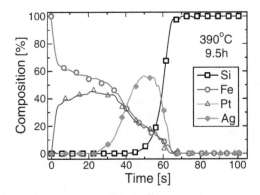

Fig. 7.10 Composition profiles of the Fe/Ag/Pt film obtained after 9.5 h at 390 °C. The symbols are for identification of the curves. (Reproduced from [23], with permission from IOP Publishing)

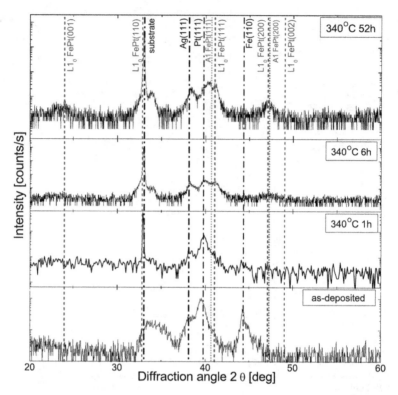

Fig. 7.11 XRD ($\theta - 2\theta$) scans obtained after annealing at 340 °C for different annealing times for the Fe/Ag/Pt film. (Reproduced from [23], with permission from IOP Publishing)

7.4 Pt/Au/Fe

The role of Au addition and behavior at low temperatures was investigated by Vladymyrskyi et al. [45]. DC magnetron sputtered Pt(15 nm)/Au(10 nm)/Fe (15 nm)/Al$_2$O$_3$ layers were annealed at 330 °C for various times. The samples were analyzed by depth profiling (SNMS), XRD and SQUID-VSM magnetometry. The transformation kinetics were compared to bilayer samples of the same Fe and Pt thickness annealed together. The first few hours of annealing resulted in diffusion of Fe through the Au grain boundaries and segregation at the Pt/Au interface (Fig. 7.13, 2 h and 4 h). Also some Fe penetration to the Pt layer could be observed. In addition Au segregation at the Fe/substrate interface was visible. After 8 h annealing these processes continued together with Pt diffusion into the Au layer (Fig. 7.13, 8 h). The penetration of Fe into the Pt layer through the Au layer (about 30 at% Fe in Au and 20 at% Fe in Pt) together with pronounced segregation of Fe at the Pt/Au interface continued after 24 h annealing (Fig. 7.13, 24 h). In addition Fe segregation at the free surface was observed and Pt appeared in the Fe layer. The processes were interpreted together with XRD patterns (Fig. 7.14), where the formation of ordered phase was

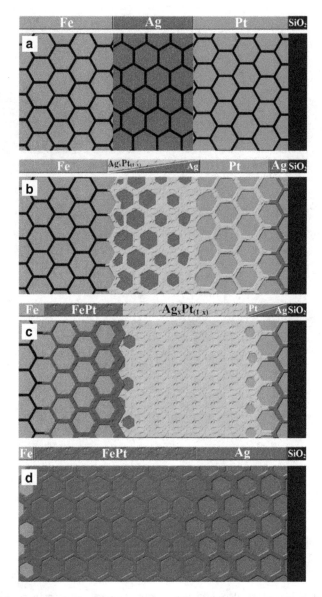

Fig. 7.12 Schematic illustration of the evolution of the solid state reaction in the Fe/Ag/Pt film sample during annealing at various temperatures: (**a**) as-deposited state (**b**) intensive intermixing between the Ag and Pt layers with the $Ag_xPt_{(1-x)}$ phase formation in Ag layer (**c**) Ag and Pt are almost fully mixed and the growth of the FePt phase has started in the GBs of Fe (**d**) last stage, when the FePt phase has been formed and the Ag has already reached the substrate with some FePt phase along its GBs. (Reproduced from [23], with permission from IOP Publishing)

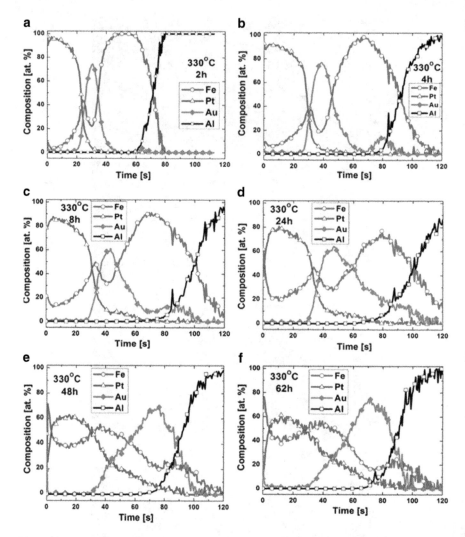

Fig. 7.13 Composition profiles of the Pt/Au/Fe films after annealing at 330 °C for (**a**) 2, (**b**) 4, (**c**) 8, (**d**) 24, (**e**) 48 and (**f**) 62 h. (Reproduced from [45], with permission from IOP Publishing)

observed after 24 h and also reflection from Au was constantly present. It was emphasized by Vladymyrskyi et al. [45] as a key finding, that no reaction layer was formed at the interfaces and also the fact, that at such low temperature bulk diffusion is practically frozen, thus only grain boundary diffusion can be taken into account. Just as in case of the previous systems the filling up of the grain boundaries can only account for ca. 10at% contamination in the layers, approximated from grain size. It was found by comparison with annealing of bilayer Fe/Pt samples (see [45] for details), that the presence of Au accelerated the diffusion process at this low

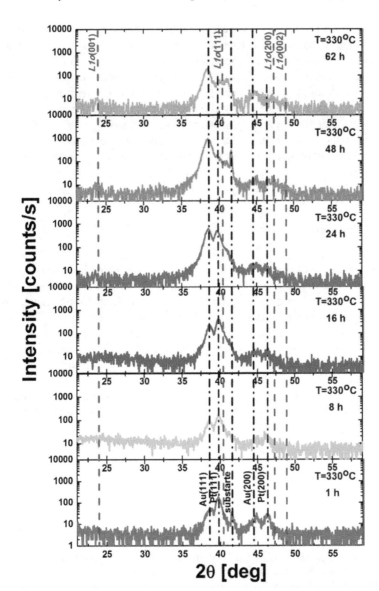

Fig. 7.14 XRD (θ–2θ) scans of Pt /Au /Fe films after annealing at 330 °C with different durations. (Reproduced from [45], with permission from IOP Publishing)

temperature and also induced the formation of ordered phase. In the final state the Au layer apparently 'moved' to the substrate, while an FePt compound is formed on top (Fig. 7.13, 48 h and 62 h).

This 'motion' of the Au layer is similar to the previously shown movement of Ag, however, the underlying mechanism was found to be different. It was emphasized, that both the Au-Pt and Au-Fe systems are almost immiscible and also the diffusion of Fe in Au grain boundaries is faster than the diffusion of Pt in Au. In this case first formation of FePt at the Au grain boundaries along with Fe diffusion into the Pt layer was assumed. Afterwards as the diffusion of Fe continued a FePt layer formed in place of the Pt layer, while the amount of FePt in the Au grain boundaries only slightly increased due to the slower diffusion. The gold layer behaved similarly to the classical inert markers in bulk interdiffusion experiments and shifted in the direction of the faster diffusing component. The more enhanced ordered phase formation in the presence of Au intermediate layer was interpreted by stress development.

7.5 Conclusion

Usually low temperature annealing is not considered as a possibility to intermix thin films and to form new, in some cases ordered phases. However, utilizing diffusion induced grain boundary migration (DIGM), which can occur during grain boundary interdiffusion, in several cases even ordered phases can be formed, which were previously only formed by high temperature annealing. Examples were reviewed in Fe-Pt-(Au, Ag) system to illustrate, that ordered $L1_0$ phases can be effectively formed by annealing below 400 °C.

Acknowledgements This work was financially supported by the GINOP-2.3.2-15-2016-00041 project co-financed by the European Union and the European Regional Development Fund.

References

1. Balluffi RW, Cahn JW (1981) Mechanism for diffusion induced grain boundary migration. Acta Metall 29:493
2. Beke DL, Szabó IA, Erdélyi Z et al (2004) Diffusion-induced stresses and their relaxation. Mater Sci Eng A 387:4
3. Beke DL, Langer GA, Molnár G et al (2013) Kinetic pathways of diffusion and solid-state reactions in nanostructured thin films. Philos Mag 93:1960
4. Blendell JE, Handwerker CA, Shen CA et al (1987) Diffusion induced interface migration in ceramics. In: Pask JA, Evans AG (eds) Ceramic microstructures '86. Springer, New York
5. Brombacher C, Grobis M, Lee J et al (2012) $L1_0$ FePtCu bit patterned media. Nanotechnology 23:025301
6. Brombacher C, Schletter H, Daniel M et al (2012) FePtCu alloy thin films: morphology, $L1_0$ chemical ordering, and perpendicular magnetic anisotropy. J Appl Phys 112:073912
7. Bulat T, Goll D (2010) Temperature dependence of the magnetic properties of $L1_0$-FePt nanostructures and films. J Appl Phys 108:113910
8. Chen C, Kitakami O, Okamoto S et al (2000) Ordering and orientation of CoPt /SiO2 granular films with additive Ag. Appl Phys Lett 76:3218
9. den Broeder FJ (1972) Interface reaction and a special form of grain diffusion in the Cr-W system. Acta Metall 20:319

10. Doherty R (1992) Grain boundary motion, diffusion-induced. In: Cahn RW (ed) Encyclopedia of materials science and engineering supplementary, vol 2. Prgamon Press, Oxford
11. Farrow RFC, Weller D, Marks RF et al (1996) Control of the axis of chemical ordering and magnetic anisotropy in epitaxial FePt films. J Appl Phys 79:5967
12. Feng C, Li BH, Liu Y et al (2008) Improvement of magnetic properties of $L1_0$-FePt film by FePt /Au multilayer structure. J Appl Phys 103:023916
13. Feng C, Zhan Q, Li B et al (2008) Magnetic properties and microstructure of FePt/Au multilayers with high perpendicular magnetocrystalline anisotropy. Appl Phys Lett 93:152513
14. Geguzin YY, Kaganovskiy YS, Paritskaya LN (1982) Cold homogenization during interdiffusion in dispersed media. Phys Met Metallogr 54:120
15. Gilbert DA, Wang LW, Klemmer TJ et al (2013) Tuning magnetic anisotropy in (001) oriented $L1_0$ $(Fe_{1-x}Cu_x)_{55}Pt_{45}$ films. Appl Phys Lett 102:132406
16. Handwerker CA (1989) Diffusion-induced grain boundary migration in thin films. In: Gupta D, Ho P (eds) Diffusion phenomena in thin films and microelectronic materials. Park Ridge, Noyes Publications
17. Handwerker CA, Cahn JW (1988) Microstructural control through diffusion-induced grain boundary migration. Mater Res Soc Symp Proc 106:127
18. Handwerker CA, Blendell JE, Interrante C et al (1993) The potential role of diffusion-induced grain-boundary migration in extended life prediction. Mater Res Soc Symp Proc 294:625
19. Hillert M, Purdy GR (1978) Chemically induced grain boundary migration. Acta Metall 26:333
20. Hsu YN, Jeong S, Laughlin DE et al (2001) Effects of Ag underlayers on the microstructure and magnetic properties of epitaxial FePt films. J Appl Phys 89:7068
21. Kai T, Maeda T, Kikitsu A et al (2004) Magnetic and electronic structures of FePtCu ternary ordered alloy. J Appl Phys 95:609
22. Katona GL, Vladymyrskyi IA, Makogon IM et al (2014) Grain boundary diffusion induced reaction layer formation in Fe/Pt thin films. Appl Phys A Mater Sci Process 115:203
23. Katona GL, Safonova NY, Ganss F (2015) Diffusion and solid state reactions in Fe/Ag/Pt and FePt/Ag thin-film systems. J Phys D Appl Phys 48:175001
24. King AH Diffusion induced grain boundary migration. Int Mater Rev 32:173
25. Kitakami O, Shimada Y, Oikawa Y et al (2001) Low-temperature ordering of $L1_0$–CoPt thin films promoted by Sn, Pb, Sb, and Bi additives. Appl Phys Lett 78:1104
26. Klinger L, Rabkin E (2011) Theory of the Kirkendall effect during grain boundary interdiffusion. Acta Mater 59:1389
27. Maeda T, Kai T, Kikitsu A et al (2000) Reduction of ordering temperature of an FePt-ordered alloy by addition of Cu. Appl Phys Lett 80:2147
28. Makarov D, Lee J, Brombacher C et al (2010) Perpendicular FePt-based exchange-coupled composite media. Appl Phys Lett 96:062501
29. Maret M, Brombacher C, Matthes P et al (2012) Anomalous x-ray diffraction measurements of long-range order in (001)-textured $L1_0$ FePtCu thin films. Phys Rev B 86:024204
30. McCallum P, Krone F, Springer C et al (2011) $L1_0$ FePt based exchange coupled composite bit patterned films. Appl Phys Lett 98:242503
31. Mittemeijer EJ, Beers AM (1978) Dislocation wall formation during interdiffusion in thin bimetallic films. Thin Solid Films 48:356
32. Mittemeijer EJ, Beers AM (1980) Recrystallization and interdiffusion in thin bimetallic films. Thin Solid Films 65:125
33. Piramanayagam SN, Srinivasan K (2009) Recording media research for future hard disk drives. J Magn Magn Mater 321:485
34. Platt CL, Wierman KW, Svedberg EB et al (2002) $L1_0$ ordering and microstructure of FePt thin films with Cu, Ag, and Au additive. J Appl Phys 92:6104
35. Rabkin E, Ma CY, Gust W (1995) Diffusion-induced grain boundary phenomena in metals and oxide ceramics. Mater Sci Monogr 81:353
36. Shamis OV, Safonova NY, Voron MM (2019) Phase transformations in Pt/Fe bilayers during post annealing probed by resistometry. J Phys Condens Mater 31:285401

37. Shewmon PG (1981) Diffusion driven grain boundary migration. Acta Metall 29:1567
38. Shewmon PG (1986) Role of moving boundaries in surface alloying. Trans Jpn Inst Met Suppl 27:443
39. Shima T, Takanashi K, Takahashi YK et al (2004) Coercivity exceeding 100 kOe in epitaxially grown FePt sputtered films. Appl Phys Lett 85:2571
40. Smith DA, Grovenor CRM (1986) Chemical effects on grain-boundray migration in Si and Ge. Trans Jpn Inst Met Suppl 27:969
41. Stephenson GB (1988) Deformation during interdiffusion. Acta Metall 36:2663
42. Tokuoka Y, Seto Y, Kato T et al (2014) Effect of Ag addition to $L1_0$ FePt and L10 FePd films grown by molecular beam epitaxy. J Appl Phys 115:17B716
43. Tu KN (1977) Kinetics of thin-film reactions between Pb and the AgPd alloy. J Appl Phys 48:3400
44. Tu KN (1985) Interdiffusion in thin films. Annu Rev Mater Sci 15:147
45. Vladymyrskyi IA, Gafarov AE, Burmak AP et al (2016) Low-temperature formation of the FePt phase in the presence of an intermediate Au layer in Pt /Au /Fe thin films. J Phys D Appl Phys 49:035003
46. Wei DH, Yuan FT, Chang HW et al (2008) Magnetization reversal and microstructure of FePt–Ag (001) particulate thin films for perpendicular magnetic recording media. J Appl Phys 103:07E116
47. Weisheit M, Schultz L, Fähler S (2007) Temperature dependence of FePt thin film growth on MgO(1 0 0). Thin Solid Films 515:3952
48. Weller D, Mosendz O, Parker G et al (2013) $L1_0$ FePtX–Y media for heat-assisted magnetic recording. Phys Status Solidi (a) 210:1245
49. Yan ML, Powers N, Sellmyer DJ (2003) Highly oriented nonepitaxially grown $L1_0$ FePt films. J Appl Phys 93:8292
50. Yoon D (1995) Theories and observations of chemically induced interface migration. Int Mater Rev 40:149
51. You CY, Takahashi YK, Hono K (2006) Particulate structure of FePt thin films enhanced by au and ag alloying. J Appl Phys 100:056105
52. Yu YS, Li HB, Li WL et al (2010) Structure and magnetic properties of magnetron-sputtered [(Fe/Pt/Fe)/Au]n multilayer films. J Magn Magn Mater 322:1770
53. Zhang L, Takahashi YK, Hono K et al (2011) $L1_0$-ordered FePtAg–C granular thin film for thermally assisted magnetic recording media. J Appl Phys 109:07B703

CPSIA information can be obtained
at www.ICGtesting.com
Printed in the USA
LVHW021254200720
661079LV00003BA/254

9 789402 420331